Accounting, Finance, Sustainability, Governance & Fraud: Theory and Application

Series Editor

Kıymet Tunca Çalıyurt, Centre for Forensic Accounting Research and Enterprise, School of Finance and Management, SOAS University of London, London, Türkiye

This Scopus indexed series acts as a forum for book publications on current research arising from debates about key topics that have emerged from global economic crises during the past several years. The importance of governance and the will to deal with corruption, fraud, and bad practice, are themes featured in volumes published in the series. These topics are not only of concern to businesses and their investors, but also to governments and supranational organizations, such as the United Nations and the European Union. Accounting, Finance, Sustainability, Governance & Fraud: Theory and Application takes on a distinctive perspective to explore crucial issues that currently have little or no coverage. Thus the series integrates both theoretical developments and practical experiences to feature themes that are topical, or are deemed to become topical within a short time. The series welcomes interdisciplinary research covering the topics of accounting, auditing, governance, and fraud.

Kıymet Tunca Çalıyurt
Editor

New Approaches to CSR, Sustainability and Accountability, Volume VI

Springer

Editor
Kıymet Tunca Çalıyurt
Business Administration and Economics
Trakya University
Edirne, Türkiye

ISSN 2509-7873 ISSN 2509-7881 (electronic)
Accounting, Finance, Sustainability, Governance & Fraud: Theory and Application
ISBN 978-981-96-5939-5 ISBN 978-981-96-5937-1 (eBook)
https://doi.org/10.1007/978-981-96-5937-1

This Springer imprint is published by the registered company Springer Nature Singapore Pte Ltd.
The registered company address is: 152 Beach Road, #21-01/04 Gateway East, Singapore 189721, Singapore

Acknowledgement

As **International Group on Governance Fraud Ethics and Corporate Social Responsibility,** we thank Prof. Achyuta Samanta for inviting us to organize the **14th International Conference on Governance Fraud Ethics and CSR** at KIIT University on November 21–22, 2024.

https://law.kiit.ac.in/14th-international-conference-on-governance-fraud-ethics-sustainability-and-reporting-icongfesr-2024/

It was also a privilege for us to be a visitor at The Kalinga Institute of Social Sciences - KISS, comprises KISS Foundation, KISS School & College and KISS University, the most prominent CSR project in the world. KISS Foundation is an NGO in India headquartered in Bhubaneswar, Odisha. School, College and University is the educational wing of this initiative located at the intersection of food, education and empowerment. KISS Foundation provide free education, accomodation, food and healthcare to over 30,000 indigenous students currently studying at our main campus in Bhubaneswar. KISS Foundation serve an additional 10,000 students across KISS satellite centers.

In our new series about KIIT and KISS, as book series editorial group, we will introduce KISS project to our readers in our upcoming books.

The Kalinga Institute of Social Sciences (KISS) comprises KISS Foundation, KISS School & College and KISS University. KISS Foundation is an NGO in India headquartered in Bhubaneswar, Odisha. School, College and University is the educational wing of this initiative located at the intersection of food, education and empowerment. We are powered by a belief that a better world—free from hunger, poverty and illiteracy—is possible. We provide free education, accommodation, food and healthcare to over 30,000 indigenous students currently studying at our main campus in Bhubaneswar. We are currently 40,000 alumni strong. We serve an additional 10,000 students across our satellite centres.

Vision

A world where all children and youth can learn free of cost to learn, lead and grow without fear to reach their full potential.

Mission

To empower 2 million indigenous children through quality comprehensive education

Odisha is known for its immense indigenous population, which comprises 24% of the total state's population. Unfortunately, these communities remain one of the most marginalized and deprived in modern Indian society. Dr. Achyuta Samanta, the founder of KIIT and KISS, realized the difficulty faced by tribal communities and attempted to empower them by founding KISS, in 1992–1993. The institute, which started with only 125 tribal students in a rented house, has spiralled into the world's largest and only institute providing free education from Kindergarten to Post-Graduation and Doctoral Programmes with vocational and extracurricular training, free lodging and boarding, state-of-the-art facilities and indigenously controlled pedagogies exclusively to more than 80,000 indigenous students. The foundations of KISS are based on Dr. Samanta's dream of providing free and accessible holistic education to indigenous students across the country. With the availability of these opportunities, students of KISS have repeatedly proven remarkable in academics, sports and vocational activities earning global recognition in both academia and sports.

https://kiss.ac.in/about/

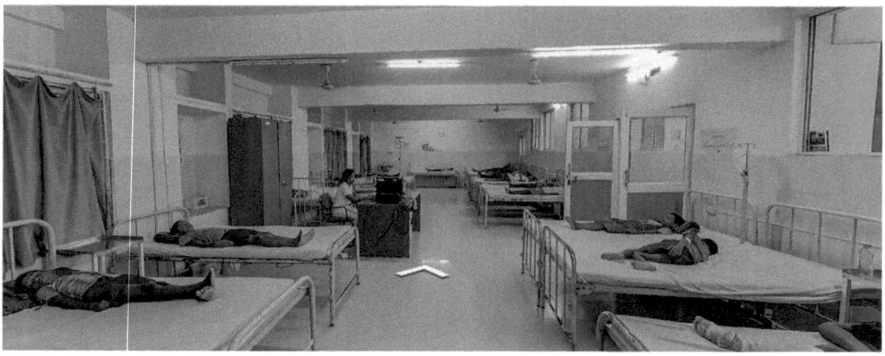

Echoes of Philanthropy, Insights from Mr. Bill Gates' interaction with my lovabel KISSians

Thank you Prof Acyuta Samanta for your support to create a better world for children. We are proud of to be part of it.

Prof. Dr. Kıymet Tunca Çalıyurt
—Founder of International Group on Governance Fraud Ethics and Corporate Social Responsibility
—President of International Conference on Governance Fraud Ethics and Corporate Social Responsibility
26.2.2025

Contents

About the Editor

Kıymet Tunca Çalıyurt graduated from Beşiktaş Atatürk Lisesi (1986) in İstanbul. Her bachelor's degree is from the Business Administration Department, Marmara University (1990) in Istanbul; her Master's (1994) and Ph.D. (2002) degrees are from the Accounting and Finance Programme at the Social Graduate School, Marmara University; her postdoctoral study is from London Metropolitan University (2004).

Prof. Çalıyurt worked as an auditor at Horwath Auditing Company, a manager at McDonald's, and finance staff at Singapore Airlines. After vast experience in the private sector, she has started working in academia. Her research interests are accounting, auditing, fraud examination, social responsibility, corporate governance, finance, business ethics, and sustainability, particularly in auditing, aviation management, NGOs, and the fast sector.

Prof. Çalıyurt was a visiting researcher at John Molson School of Business in 2001 summer (International Centre for Aviation Management), Massachusetts University Amherst Business School in the 2014 summer, the 2018 summer, 2024 summer (Accounting and Information System), University of National World and Economy in the 2019 summer (Faculty of Accounting). She founded the International Group on Governance, Fraud, Ethics and Social Responsibility (IGonGFE&SR) and the International Women and Business Group (IWBG) in 2009; both groups organized annual conferences and published international book series.

Prof. Çalıyurt performs as

- Book series editor in Springer indexed in Scopus, titled "Accounting, Finance, Sustainability, Governance & Fraud: Theory and Application" [1], and in Routledge titled "Women and Sustainable Business" [2].
- Attends as an invited speaker at the Cambridge Economic Crime Symposium and a workshop organizer on the role of external auditors in fraud prevention.
- Editorial board member in the Journal of Financial Crime (Web of Science - ESCI), International Journal of Law and Management (Web of Science - ESCI), Journal of Money Laundering Control (Web of Science—ESCI), International Journal of Climate Change Strategies and Management (Web of Science—SSCI).

She is a founding partner of Herme Consulting, located in Trakya University Technopark. The company trains and organizes company fraud examinations, sustainability, auditing, accountability and academic publishing.

List of Figures

List of Tables

List of Graphs

Part I
Introduction

Chapter 1
Introduction: Sustainability Reporting in Big Four

Aslı Selvi and Kıymet Tunca Çalıyurt

Abstract Current events are transforming the way we live and do business. CSRD (Corporate Sustainability Reporting Directive) auditing involves an independent, external review of a company's sustainability report which requires assurance by an external auditor. Auditing the sustainability statement involves an independent, external review of a company's sustainability report. This process ensures that the information disclosed is accurate, complete, and in accordance with the European Sustainability Reporting Standards (ESRS). The audit offers stakeholders confidence in the reported data, which is crucial for informed decision-making (Ekobiomanager, CSRD auditing: how and why you need to audit?, 2025). Our aim of this study is to examine the content of the "self-published sustainability reports" of the four major audit companies authorized to audit sustainability reporting. The absence of previous research on this subject shows the study's originality.

Keywords Auditing · Auditing firm · Sustainability reporting · Big four

A. Selvi
Trakya University, Edirne, Türkiye

Social Graduate School, Edirne, Türkiye
e-mail: asliselvi3@trakya.edu.tr

K. T. Çalıyurt (✉)
Business Administration Department, Faculty of Business Administration and Economics, Trakya University, Edirne, Türkiye

Accounting and Information Systems Department, CARLAB, Business School, Rutgers University, New Jersey, USA

K. T. Çalıyurt (ed.), *New Approaches to CSR, Sustainability and Accountability, Volume VI*, Accounting, Finance, Sustainability, Governance & Fraud: Theory and Application, https://doi.org/10.1007/978-981-96-5937-1_1

Abbreviations

CAQ	Center of Audit Quality
CSRD	Corporate Sustainability Reporting Directive
ESG	Environmental, Social and Governance
EY	Ernst & Young
FY	Financial Year
IAASB	International Auditing and Assurance Standards Board
ISO	International Standardization Organization
Kwh	Kilowatt-hour
m^3	Cubic meter
PCAOB	The Public Company Accounting Oversight Board
PwC	Price waterhouse coopers
SASB	Sustainability Accounting Standards Board
SEC	Securities and Exchange Commission

1.1 Introduction

There is no widely accepted definition of sustainability reporting. However, some institutions have defined this. One of these is done by Global Reporting Initiative (GRI) as sustainability reporting is a process that helps businesses set accurate and acceptable goals for a sustainable global economy, measure performance and manage change, while combining long-term profitability with social responsibility and environmental care. In addition, it is an important platform element to communicate the organization's economic, environmental, social, and governance (ESG) performance and reflect its positive and negative impacts with sustainability reporting (Global Reporting Initiative 2013). Sustainability reporting is preferred for business continuity reasons, which include enhancing corporate reputation and transparency, strengthening competitiveness, managing risks, making improvements in the environmental, social, and economic context, providing more reliable and higher quality information to shareholders and increasing share value through this information (Akarçay 2014). When investment experts and other stakeholders are evaluated, it is a known fact that the long-term growth of companies does not depend solely on financial profitability. Therefore, companies have turned to sustainability reporting, which helps increase corporate transparency regarding environmental and social issues beyond the narrow economic perspective. Companies assume that sustainability reporting is costly, but according to research, there is a relationship between companies publishing sustainability reports and firm value (Berthelot et al. 2012; Kuzey et al. 2023). The purpose of sustainability reporting is to increase transparency, enhance brand value, meet legal obligations, enable mutual assessment of competitors, increase employee motivation, and support and control corporate knowledge (Hyršlová 2014).

Guidry and Patten (2010) found that firms publishing high-quality sustainability reports receive significantly more positive market responses than firms publishing lower-quality sustainability reports. Environmental, social, and governance (ESG) reports help firms mitigate the negative valuation effects of negative ESG performance (Fatemi et al. 2018). Businesses that adopt sustainability reporting will be able to identify potential opportunities and risks in advance, allowing them to manage proactively and efficiently, and adapt more quickly to new conditions. Lastly, with the reporting practices related to sustainability performance, governments will be able to make more accurate predictions about the impact and contribution of businesses to the economy (Ozata Canli and Serçemeli 2024).

1.2 Sustainability Reporting by Audit Companies

When reviewing the published reports, it is evident that the Securities and Exchange Commission (SEC) and the Public Company Accounting Oversight Board (PCAOB) have shown increasing interest in sustainability disclosures and the role of the auditor in sustainability reporting. Initially, in 2010, the SEC provided guidance on existing climate change risk disclosure requirements (Lee 2021). In March 2022, the SEC proposed significant rule changes requiring firms to include climate-related disclosures, such as sustainability risks and related risk management processes, risk materiality and impact, greenhouse gas emissions, and climate-related events (SEC 2022).

It is stated that auditors are ready to close the gap between sustainability guidance under existing and proposed frameworks and sustainability reporting of client firms. Julie Bell Lindsay, Executive Director of the Center for Audit Quality (CAQ), states that auditors bring the necessary independence, expertise, and experience to enhance the reliability of ESG reporting, as this information plays an increasingly important role in investment strategies (Liu et al. 2023). According to another study by Pflugrath et al. (2011), sustainability reporting produces more reliable results when provided by professional accountants rather than sustainability consultants. Auditors and affiliated firms play a key role in sustainability reporting (O'Dwyer 2011) and provide 63% of global assurance in sustainability reporting (International Federation of Accountants 2021). As sustainability becomes a strategic issue for firms as climate change becomes more critical, auditors can facilitate the role of accounting in assessing the environmental, social, and governance (ESG) outcomes and broader societal impacts of a firm's operations (Liu et al. 2023).

Post-audit financial reporting quality is measured using earnings management structures calculated based on financial reports published after the audit is completed (LópezPuertas-Lamy et al. 2017). The evidence also correlates positively with the quality of sustainability reporting. The evidence obtained here can also be applied to sustainability reporting (Al-Shaer and Zaman 2019).

In recent years, many firms have successfully sought to develop and secure new markets in areas such as quality auditing (Power 1997a, b; Swift et al. 2000) and environmental auditing (Power 1997a, b, 2003). These changes have been partly facilitated by the ability of accountants to translate the concepts and terminology underpinning attestation audits from their traditional methods into non-financial audit areas (Free et al. 2009; Power 1997a, b).

Fig. 1.1 Logos of Big Four. Source: Mckenna (2017)

The economic success of the Big 4 accounting firms appears to be influenced by their traditional bases in audit, tax, and insolvency, as well as their willingness and ability to expand into new and potentially lucrative areas of business (O'Dwyer 2011) (Fig. 1.1).

The Big 4 Audit Company (KPMG, Deloitte, Ernst & Young, and PwC) have a significant share of the growing market for assurance on sustainability reports. This market is shared between certification bodies, specialist consultancies, and the Big 4 professional services firms. According to a global study conducted by KPMG in 2008, the Big 4 control 70% of the assurance market among G250 companies. As a result of this growth and the increased involvement of the accounting profession in its provision, sustainability assurance is on the agenda of many international and national accounting standard setters. These include the International Auditing and Assurance Standards Board (IAASB), which has stated its intention to develop a sustainability assurance standard as one of its strategic objectives for 2009–2011 (IAASB 2007).

According to Ernst & Young's sustainability report published in 2022, innovation is at the core of a future-oriented business. All citizens in a society are considered to have a valuable competence because they can achieve unique solutions together with different perspectives. Ernst & Young argues that the challenges of our time depend on our ability to understand the competence of others and how we can contribute to a better future (Ernst Young, 2023a).

Auditing firms are the companies they provide "trustable economic environment" in the business as good citizens.

Audit companies have the authority to prepare reports to ensure that financial statements in the world economy are transparent and accountable in accordance with accounting and auditing standards. In this context, audit companies must explain their experiences with environmental-social-governance while doing business. To achieve this, audit companies have started to publish sustainability reports.

1.3 Methodology

In this study, the "Content Analysis" method is used. Along with the content analysis, the "Sustainability Report" published by PwC in Turkiye in 2023 is examined. This report was also supported by other publications. "Content Analysis" is described as the scientific study of content of communication (Prasad 2008).

In 1952, Bernard Berelson introduced the technique of content analysis as a versatile tool for social science and media researchers. This method, which was not immensely popular in its early years, gained serious popularity among social science academics after a certain period (Wimmer and Dominick, 1994). Since the 2000s, technology has radically changed the way people communicate with each other. The proliferation of email, the web, digital photography, social media, YouTube, text messaging, and mobile phones has produced an unprecedented amount of permanent, archived data about individuals (Stemler 2015).

The resources provided for the research and the methodological advances made in the context of the problems examined within the scope of these resources have contributed significantly to the emergence of the methodology in this method.

There are three basic principles of content analysis. These can be summarized as follows:

1. Objectivity: This means that the analysis is conducted based on clear rules that ensure that different researchers will obtain the same results from the same documents or messages.
2. Systematic: The inclusion or exclusion of content is done according to some consistently applied rules that exclude the possibility of including only material that supports the researcher's ideas.
3. Generalizability: The results obtained by the researcher can be applied to other similar situations (Prasad 2008).

As a result, analysts have dubbed this the era of "big data." Both private companies and public government organizations are actively trying to extract this data to discover individual and group behavioral patterns. This data will be used to conduct analyses in this area. However, in order to fully utilize big data, appropriate methods for data analysis must be used. As a result, content analysis as a methodological technique has yielded positive results in this area and is also applied to big data. Content analysis can be used with a wide variety of data sources, including textual data, visual data (e.g., photos/videos), and audio data. In addition, the technique is widely used in the social sciences because it is highly flexible in terms of being empirically or theoretically driven (Stemler 2015).

In this study, content analysis is applied on texts. There are 4 major auditing companies in the world. These are PwC, Deloitte, KPMG, and Ernst & Young. In this study, PwC Turkey's 2023 Sustainability Report is examined and supported by reports from other auditing companies.

Fig. 1.2 The service structure of the PwC ESG Platform. Source: PwC Turkiye Sustainability Report, 2023

1.4 Sustainability Report Analysis of Auditing Companies

This section is based on the "Sustainability Report" published by PwC Turkiye in 2023. In addition to this report, reports from auditing companies Deloitte, Ernst & Young, and KPMG were also used (KPMG Sustainability Report 2019–2021 Türkiye).

The service structure of PwC's ESG Platform can be grouped under three headings (Fig. 1.2).

Strategy: PwC's future clients are helped to make the right choices to position them for a more sustainable and inclusive tomorrow.

Reporting and Audit: We support our customer to explain their successfull sustainability transformation in the most transparent and the accurate way.

Transformation: We support clients in integrating an environmental, social, and governance (ESG) perspective into their daily operations and business practices (PwC Turkiye Sustainability Report, 2023).

Sustainability can be shaped around four main focus areas.

Responsible Service Approach: As a professional service provider, we work to build trust in society and solve important problems. It is significant for organizations to reduce the risks arising from their activities, seize opportunities, and create more sustainable value. Responsible service approach is recognized as a part of ethics and compliance, quality, risk management, data security and privacy. In this respect, it is particularly important that client firms provide audit companies with an important level of trust and receive the same level of client satisfaction feedback. This situation has a serious impact on client satisfaction as well as audit quality. In addition to this understanding, digital transformation, which has entered our lives more widely with the Covid-19 pandemic, is taken as a guide.

Environment: When we understand the impact of unnecessary social units on the environment, the economic planet, and the environment, we need to start working on urgent issues such as climate change. Everything we have done and are doing, including our impact on our planet, should aim to build trust in society and solve important problems. It is particularly important to support the visibility of the largest life of society and the technology that covers a large part of our daily lives and the society/workers to achieve more sustainable results. The increasing global warming with climate change will have profoundly serious consequences for the environment. For this, it is necessary to show the change of all social parts.

Employees: Intentionality also means working with others to address the biggest challenges facing the world in a trust-building way. That's why we believe it's our responsibility to be part of the solution to societal problems and help people adapt to the future.

Society: The employee ecosystem provides the foundation for the ability to continue to provide high-quality services that customers expect from companies. Therefore, it is extremely important for employees in companies to value and operate with this awareness. The HR strategies of companies are focused on being the world's leading talent developer and providing employees with more agility and confidence in a rapidly changing world. For employees, their work, knowing this, adds value to both the company they work for and the customer companies (PwC Turkiye Sustainability Report, 2023).

In FY22, 10 primary (high) priority and 6 secondary priority issues are included in the matrix around 4 strategic themes consisting of "Responsible Service Approach," "Environment," "Employees," and "Society." The prioritization matrix is a representation of the primary priority issues (and their relative priorities) for the long-term strategy. The priority issues included here are the issues that PwC Turkey's stakeholders believe should be addressed in the audit companies' strategy. The determination of the determined priority issues was carried out through an online survey. A stakeholder analysis was also conducted from the results obtained. The results obtained were shared with the senior management and a final result was obtained. The results are shown in Fig. 1.3. The results obtained when comparing Stakeholder Priorities and Company Strategy Priorities are shown in Fig. 1.2. While variables numbered 15 and 16 represent the lowest level of importance, variables such as 1 and 2 represent the highest level of importance. This figure provides us with a brief summary of which variables have the highest level of importance in creating value in the long term, such as PwC Turkiye (Fig. 1.4).

At the same time, PwC aims to accelerate the transition to a low-carbon economy by committing to the United Nations' Race to Zero campaign and the Business Ambition for 1.5 °C. There are three main action plans within the scope of the 2030 Net-Zero Commitment. These can be summarized as follows:

Increasing Efficiency: Preventing and reducing carbon emissions from activities is accepted as the starting point of the energy reduction strategy.

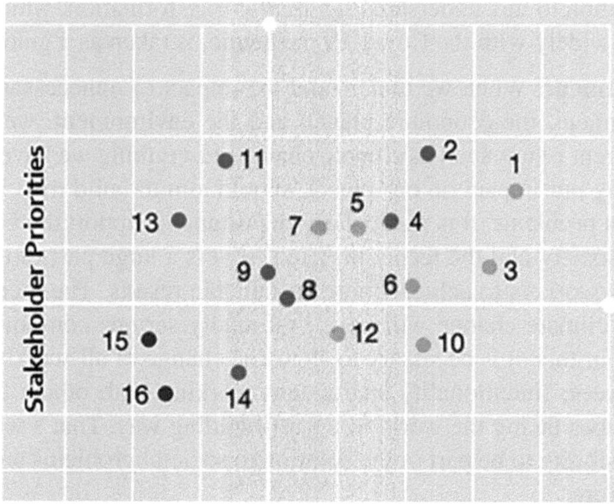

Fig. 1.3 The Materiality Matrix. *1.Ethics and Compliance, 2.Climate Change, 3.Service Quality, 4.Sustainability in the Value Chain, 5.Data Security and Privacy, 6.Risk Management, 7.Digital Transformation, 8.Inclusion and Diversity, 9.Employee Health and Welfare, 10.Public Trust and Reputation, 11.Waste Management, 12.Thought Leadership, 13.Human Rights, 14.Talent Management and Recruitment, 15.Employment and Economic Contribution, 16.Positive Impact on Society and Environment. Source: PwC Turkiye Sustainability Report, 2023

Transition to Renewable Resources: The use of renewable energy resources is considered to be one of the most effective ways to reduce environmental impacts.

Carbon Offset: Carbon offset projects are supported to minimize climate impacts. As PwC Turkey, we adopt the principle of continuous development in order to comply with the decisions made in the global context. Work continues to zero out carbon emissions created on the environment by 2030. Work is also carried out to increase efficiency, carbon footprint calculation and mapping are carried out, and the work on the transition to renewable energy continues.

The "+" signs in the boxes in Table 1.1 mean that the audit companies are included in their sustainability reports. The empty parts mean that they are not included in that audit company. When the results are examined, it is seen that all/ most of the important factors are included in the audit company, while some definitions are not included. Audit companies prepare sustainability reports according to their own requests without taking a specific format as a basis. Therefore, certain differences occur. However, basically, both country-wide and holistic sustainability reports meet the required answers. The publication of this report by all audit companies, not just large audit companies, will produce positive results for stakeholders.

Since 2019, PwC has been calculating its carbon footprint based on the methodology of the Greenhouse Gas Protocol (GHG Protocol). In 2030, annual reduction targets are determined, and actions are taken accordingly thanks to the scenario analysis carried out with the target of net-zero commitment (Graph 1.1).

In the Financial Year 2022, carbon emissions increased 1.65 times compared to the last reporting period. The main reason for this is the easing of travel restrictions imposed during the COVID-19 pandemic and our employees' hybrid working

As a RE100 participant, all PwC offices use 100% electricity generated from renewable energy sources,

Integrating the impacts of climate change and other environmental, social and governance (ESG) issues into our business as part of the transformation,

Improving non-financial reporting so stakeholders can better understand the business's climate impact,

Reshaping the service model offered to customers by balancing remote and office work,

At least 50% of product and service suppliers will set science-based targets by 2025,

Elimination of emissions resulting from activities through carbon offset projects as of 2030,

Achieving a 50% reduction in total carbon emissions from our activities by 2030, taking 2019 as the base year,

Providing thought and discussion environments on how PwC, as an auditing firm, can help reform the structures of the economy in terms of climate dimension,

It is aimed at increasing energy efficiency by carrying out improvement works in offices.

Fig. 1.4 PwC Turkey's 2030 Net-Zero commitment targets. (PwC Turkiye Sustainability Report, 2023)

model, allowing them to work in the office on certain working days of the month. After the pandemic, carbon emissions, which increased with the return to work, are expected to decrease linearly in the coming periods in line with our 2030 net-zero target. To this end, we are investing in energy efficiency improvement projects in buildings and accelerating our renewable energy transformation. At the same time, energy efficiency is considered and supported at every stage of our operations, from product and service procurement to all our operations (Graph 1.2).

Nine out of 10 respondents to the EY survey have made public climate change commitments, a higher rate than Fortune Global 500 companies. However, there is still room for improvement. These can be summarized as follows:

- Only 19% of companies are carbon negative and 12% have made net-zero carbon commitments.
- Most companies are falling short of their target of a 45% reduction by 2030 to achieve 1.5 °C (as set out in the Paris Agreement).
- 55% of companies have a reduction commitment by 2030.
- Only 27% of companies plan to reduce their emissions by 45% or more (EY Turkiye Sustainable Value Study, 2023a).

Table 1.1 Comparison of Big 4s' sustainability reporting

Sustainability report headings	Audit company names			
	Deloitte	KPMG	PwC	Ernst & Young
Company general structure		+	+	+
Note from the leader	+	+	+	+
Notifications from stakeholders	+			
Sustainability approach			+	
Responsible service approach			+	
Environment			+	
Employee			+	
Client				+
People				+
Society			+	+
ESG journey		+	+	
Intellectual capital		+		
Financial capital		+		
Natural capital		+		
Human capital		+		
Social and relationship capital		+		
Carbon neutral certificates		+		
Methodology	+			
GRI index				+
Assurance report				+
Studies on climate change	+			
Recommendations	+			
Appendices		+	+	+

Table 1.1 created by the author (Deloitte Global, 2024; KPMG Turkiye, 2019–2021; PwC Turkiye, 2023; Ernst & Young Switzerland, 2023a)

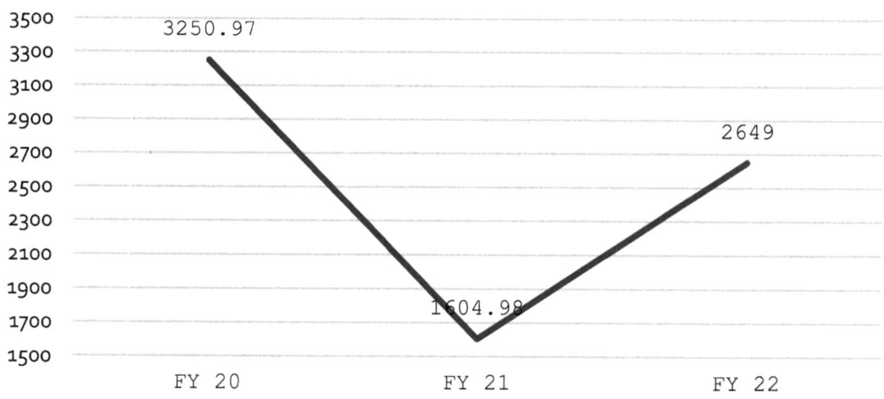

Graph 1.1 Total emissions. Source: PwC Turkiye Sustainability Report, 2023

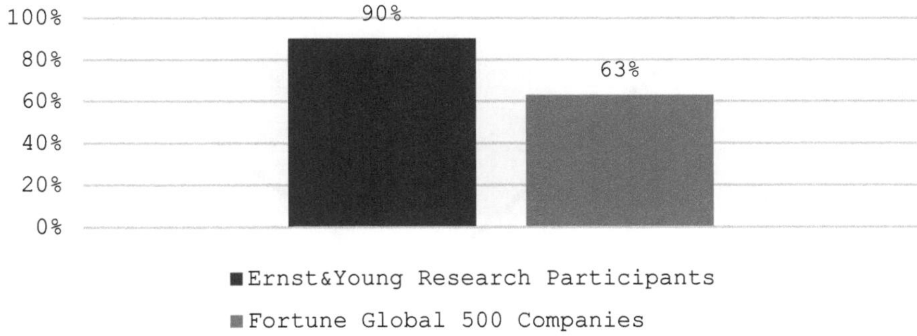

Graph 1.2 Ratio of publicly disclosed climate change commitments. Ernst Young, 2023c

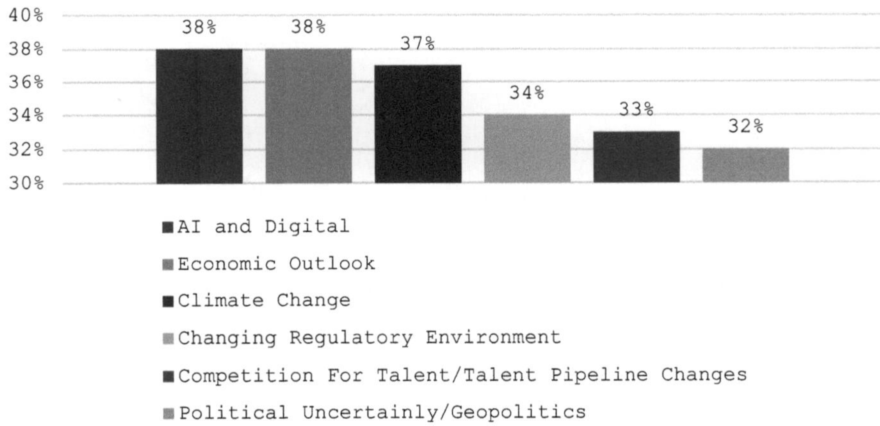

Graph 1.3 What does your organization see as the most pressing issues to focus on over the next year? *Source*: Deloitte Global Sustainability Report, 2024

According to the EY Sustainable Value Study, prepared with the participation of senior executives responsible for sustainability in Turkey, only 27% of companies plan to reduce their emissions by 45% or more, and only 55% have committed to reducing emissions by 2030. Turkish companies are ahead of global companies in making climate commitments, but they are not as fast in fulfilling these commitments and taking action. 90% of the companies participating in the study have made public climate change commitments. On average, companies plan to reduce their emissions by 34%, but it is noted that only 23% have been reduced so far (Ernst Young, 2023b) (Graph 1.3).

According to research conducted by Deloitte Global, climate change is one of the issues that companies should give most importance to (Graph 1.4).

Major changes continue to be made in the way we generate and use electricity to minimize the effects of climate change caused by human activities. For this reason, the use of renewable energy sources, which is a sustainable and environmentally

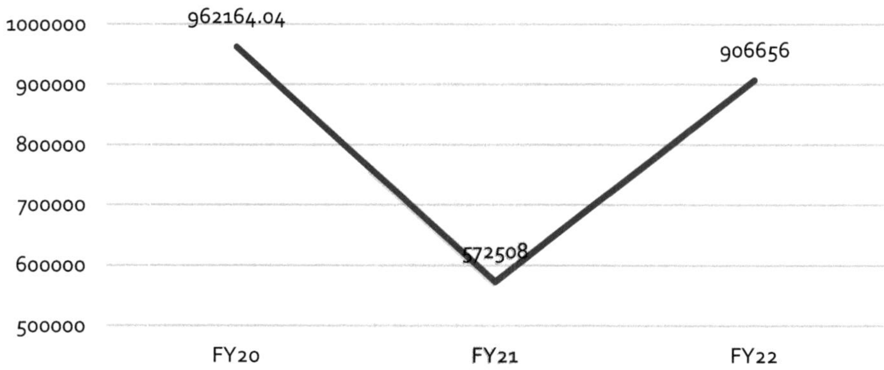

Graph 1.4 Electricity consumption in our buildings (kWh). *Source*: PwC Turkiye Sustainability Report, 2023

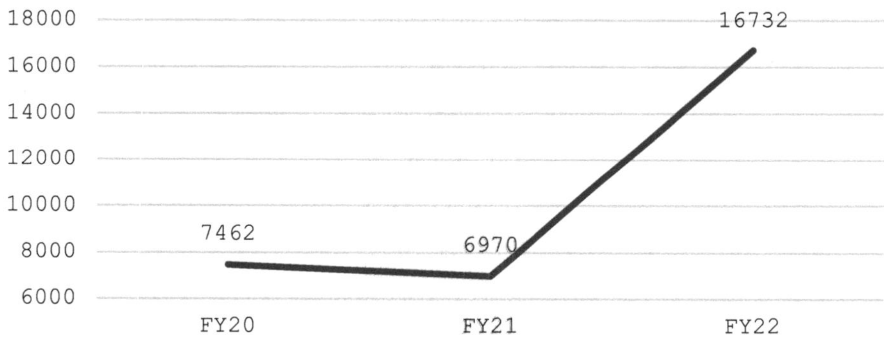

Graph 1.5 Water consumption (m^3). *Source*: PwC Turkiye Sustainability Report, 2023

friendly approach that helps reduce greenhouse gas emissions, is seen as one of the most effective solutions. PwC prioritizes the purchase of electricity generated from renewable energy sources in offices to reduce the environmental impact of operations. As part of our commitment to be 100% renewable, in 2018, PwC joined RE100, an influential global coalition of businesses committed to using 100% renewable energy (Graph 1.5).

Freshwater supply has been directly affected by a global temperature increase of ~1 °C over the last half century, resulting in extreme precipitation, more frequent floods and droughts, rising sea levels, and changing rainfall regimes. If current trends continue, nearly half (47%) of the world's population is projected to live in areas of high-water scarcity by 2030. With the start of post-Covid returns to work, there is an increase in water consumption in FY22 compared to FY20 and FY21.

With the excessive use of water in today's world, both individuals and the largest organizations need to be more aware in this field to prevent the extreme drought that will occur in the coming years. Considering that there cannot be a life without water is also very important for later generations (Graph 1.6).

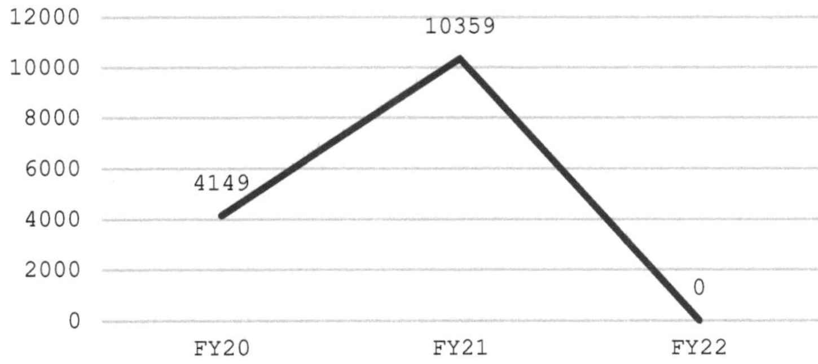

Graph 1.6 Single-use plastic consumption - pieces. *Source*: PwC Turkiye Sustainability Report, 2023

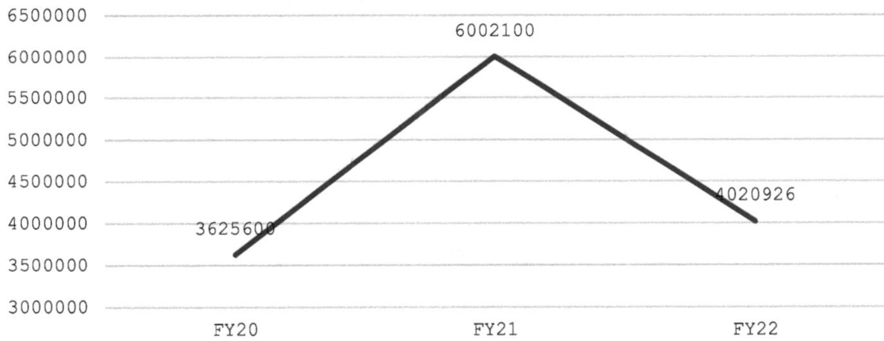

Graph 1.7 Paper consumption—pieces. *Source*: PwC Turkiye Sustainability Report, 2023

The waste generated because of PwC's activities consists of paper, electronic, and household waste used in offices. Although PwC's business is not a production activity, waste management is one of its priorities to minimize environmental impact and contribute to the circular economy (Graph 1.7).

In FY22, PwC Turkey reduced paper use in offices by 33% compared to the previous year. We aim to significantly reduce the use of paper used in offices, thereby encouraging more sustainable office operations.

Reducing paper consumption is particularly important directly for our environment. Hundreds of trees are cut down for the paper used, which affects climate change. When the "Audit Sector" is taken as a basis with digitalization, the preparation of both working papers and other studies in computer environment and storing them in digital environments will reduce the level of paper use (Graph 1.8).

Recycling and reuse are two of the most important topics of sustainability. Both PwC and the community support efforts in this area. Reductions in production costs through recycling and reuse are also of great economic importance.

On a company basis, one of the most important sustainability issues is employee sustainability(Graph 1.9). A high turnover rate, such as in the audit sector, also affects the quality of the audit (Graph 1.10). Therefore, it is a critical issue to ensure

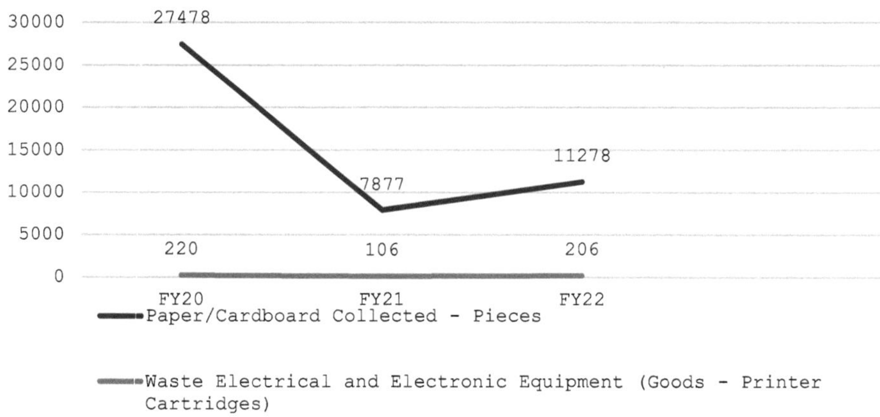

Graph 1.8 Recycling/Reuse. *Source*: PwC Turkiye Sustainability Report, 2023

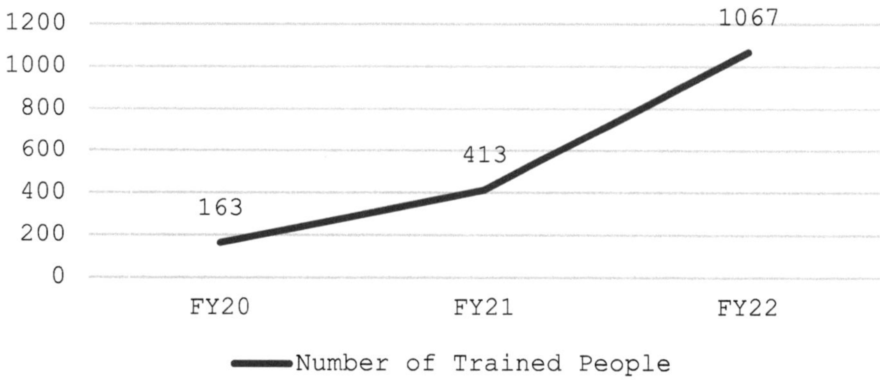

Graph 1.9 Number of people trained as part of the Digital Academy

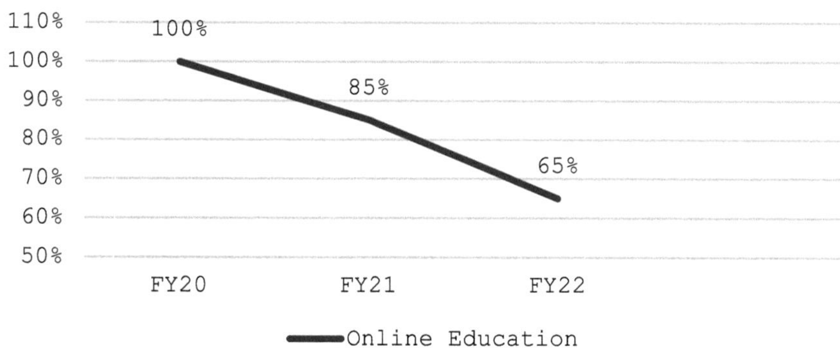

Graph 1.10 Share of online trainings

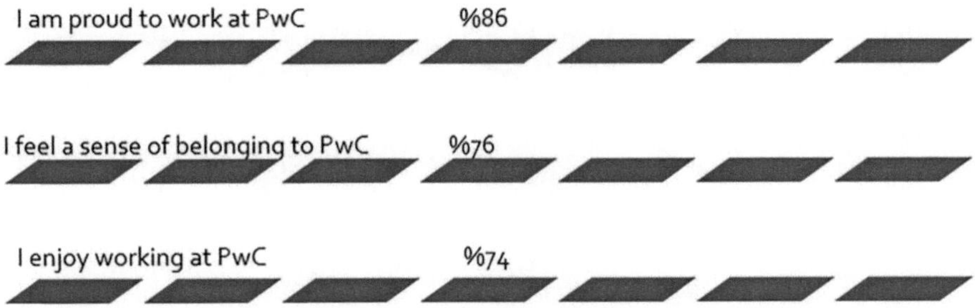

Fig. 1.5 Employee satisfaction rates. *Source*: PwC Turkiye Sustainability Report, 2023

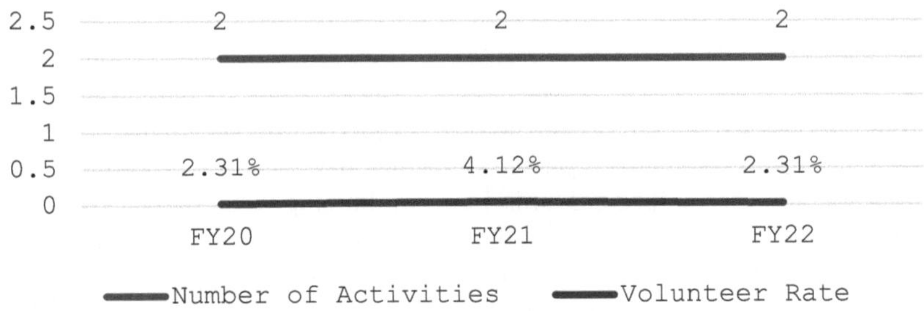

Graph 1.11 Volunteer working

the continuity of the people working in audit companies (Fig. 1.5). When we look at the satisfaction survey conducted by PwC Turkiye;

There has been a consistent increase in the number of people receiving training through the digital academy. However, when looking at online training, the rate, which reached 100% during the Covid-19 period, dropped to 65% by 2022. With the return to normalcy after Covid-19, the share of in-person training has increased, and methods based on peer learning have been prioritized in these trainings, leading to a decline in the share of online trainings (Graph 1.11). However, when considering the area of training, online training continues to be offered (Graph 1.12).

In PwC Turkiye, activities and the rate of volunteer work need to increase in volunteer work. When looking at the FY20, FY21, and FY22 rates, there is no change in the number of activities, while the number of varieties is low. Increases in the number of activities in line with the requests of employees will positively affect these rates. Activities that may occur on the Environment and Society side may also have positive results in terms of climate change.

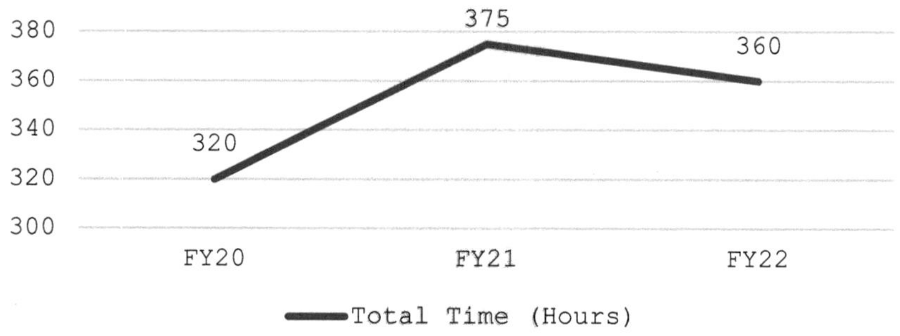

Graph 1.12 Volunteer working hours

1.5 Conclusion

Sustainability has gained importance in recent years as climate change has reached more critical levels. Both society and companies have much to contribute in this regard. Reducing greenhouse gas emissions, minimizing water and electricity consumption, and transitioning to renewable energy systems in electricity usage are of great importance for both the ecosystem and sustainability.

From the perspective of the auditing industry, sustainability reporting and disclosure draw on various sources, including SEC, SASB, GRI, and ISO, making it challenging for clients to fully integrate sustainability information into their reporting practices. This situation can lead to a constructive critical perspective between auditors focused on sustainability and clients aiming to enhance their sustainability reporting. It has been observed that the auditor's focus on sustainability is positively related to the client's sustainability reporting. Although the cost of sustainability reporting can be high, it is crucial for the company publishing the report, the client, and stakeholders in terms of commitments made and future planning. In today's world, where the global ecosystem faces increasingly difficult challenges, it is important to monitor the measures companies take through their published reports.

This study focuses on the sustainability reports of audit companies. A content analysis was conducted based on PwC Turkey's 2023 Sustainability Analysis. According to the results of the study, it was found that there are some issues in Turkey regarding companies' progress toward achieving the Net-Zero carbon emission targets they have committed to. While companies in Turkey are faster at signing agreements, they appear to be far from achieving these targets. More stringent policies need to be implemented to ensure sustainability and stay on track with these goals. Since we only have one planet and resources are not unlimited, everyone should act accordingly and take appropriate actions.

When the sustainability reports of the four major audit companies were examined, it was observed that each firm addressed different topics. These topics are summarized in Table 1.1. This study is unique in its field, as it examines the commitments and actions of the four major audit companies.

References

Akarçay Ç (2014) Sürdürülebilirlik Muhasebesi Standartları Kurulu-Sustainability Accounting Standards Board. Öneri Dergisi 11(42):1–11

Al-Shaer H, Zaman M (2019) CEO compensation and sustainability reporting assurance: evidence from the UK. J Bus Ethics 158:233–252

Berthelot S, Coulmont M, Serret V (2012) Do investors value sustainability reports? A Canadian study. Corp Soc Respon Environ Manag 19(6):355–363

Ernst Young (2023a) How Can Value Realized Today Reshape Tomorrow?, EY Switzerland Sustainability Report (2023) https://assets.ey.com/content/dam/ey-sites/ey-com/en_ch/noindex/ey-switzerlandsustainability-report-switzerland-12-2023.pdf. Accessed 21 Oct 2024

Ernst Young (2023b) EY CESA Sustainable Value Study: Six in ten companies plan to spend more money next year to address climate change, https://www.ey.com/en_ro/newsroom/2023/07/ey-cesa-sustainable-value-study-six-in-ten-companies-plan-to-spend-more-money, (Accessed on March 16th, 2024)

Ernst Young (2023c) The world's top 500 family businesses include 16 family companies from Switzerland, https://www.ey.com/content/dam/ey-unified-site/ey-com/en-ch/newsroom/2023/1/documents/ey-family-business-index-2023-switzerland-en.pdf. (Accessed on March 16th, 2024)

Deloitte Global (2024) CxO sustainability report signs of a shift in business climate action, https://www.deloitte.com/content/dam/assets-shared/docs/about/2024/deloitte-2024-cxo-sustainability-report.pdf. Accessed 16 Oct 2024

Fatemi A, Glaum M, Kaiser S (2018) ESG performance and firm value: the moderating role of disclosure. Glob Financ J 38:45–64

Free C, Salterio SE, Shearer T (2009) The construction of auditability: MBA rankings and assurance in practice. Account Org Soc 34(1):119–140

Global Reporting Initiative (2013), GRI G4 sustainability reporting guidelines. https://respect.international/wp-content/uploads/2017/10/G4-Sustainability-Reporting-Guidelines-Implementation-Manual-GRI-2013.pdf. Accessed 20 Oct 2024

Guidry RP, Patten DM (2010) Market reactions to the first-time issuance of corporate sustainability reports: evidence that quality matters. Sustain Account Manag Policy J 1(1):33–50

Hyršlová J (2014) Sustainability reporting—a review and trends and situation in CR. In The 8th international days of statistics and economics, conference proceedings, Slaný, Libuše Macáková, Melandrium, pp 497–506

IAASB (International Auditing and Assurance Standards Board) (2007) Proposed Strategy for 2009-2011. International Federation of Accountants (IFAC), New York

International Federation of Accountants (2021) The state of play in sustainability assurance | Benchmarking global practice. https://www.ifac.org/system/files/publications/files/IFAC-Benchmarking-Global-Practice-Sustainability-Assurance.pdf. Accessed 20 Oct 2024

KPMG Sustainability Report 2019–2021 Turkiye. https://assets.kpmg.com/content/dam/kpmg/tr/pdf/2022/07/sustainability-report.pdf. Accessed 18 Oct 2024.

Kuzey C, Elbardan H, Uyar A, Karaman AS (2023) Do shareholders appreciate the audit committee and auditor moderation? Evidence from sustainability reporting. Int J Account Inform Manag 31(5):808–837

Lee AH (2021) Statement on the review of climate-related disclosure. SEC statement. https://www.sec.gov/news/public-statement/lee-statement-review-climate-related-disclosure. Accessed 21 Oct 2024

Liu M, Tang J, Walton S, Zhang Y, Zhao X (2023) Auditor sustainability focus and client sustainability reporting. Account Org Soc:101512

LópezPuertas-Lamy M, Desender K, Epure M (2017) Corporate social responsibility and the assessment by auditors of the risk of material misstatement. J Bus Financ Acc 44(9–10):1276–1314

Mckenna F, (2017) Big four audit firms enjoy a "Too Few to Fail" regulatory Hall Pass. https://www.promarket.org/2017/07/14/big-four-audit-firms-enjoy-fail-regulatory-hall-pass/. Accessed 15 Nov 2024
O'Dwyer B (2011) The case of sustainability assurance: constructing a new assurance service. Contemp Account Res 28(4):1230–1266
Ozata Canli SN, Serçemeli M (2024) Sürdürülebilirlik Raporlamasında Önemlilik Değerlendirme Süreci: BIST Şirketleri Üzerine Ampirik Çalişma, Financial Analysis/Mali Cozum Dergisi, 2024, 34(181):125
Pflugrath G, Roebuck P, Simnett R (2011) Impact of assurance and assurer's professional affiliation on financial analysts' assessment of credibility of corporate social responsibility information. Auditing J Pract Theory 30(3):239–254
Prasad BD (2008) Content analysis. Res Meth Soc Work 5(le20)
Power M (1997a) Expertise and the construction of relevance: accountants and environmental audit. Account Org Soc 22(2):123–146
Power M (1997b) The audit society: Rituals of verification. OUP, Oxford
Power MK (2003) Auditing and the production of legitimacy. Account Org Soc 28(4):379–394
PwC Sustainability Report (2023) Turkiye, https://www.pwc.com.tr/tr/hakkimizda/assets/pwc-turkiye-fy22-surdurulebilirlik-raporu.pdf. Accessed 19 Oct 2024
Securities and Exchange Commission (SEC) (2022) SEC proposes rules to enhance and standardize climate-related disclosures for investors. https://www.sec.gov/news/press-release/2022-46. Accessed 21 Oct 2024
Stemler SE (2015) Content analysis. Emerging trends in the social and behavioral sciences: an interdisciplinary, searchable, and linkable resource, pp 1–14
Swift TA, Humphrey C, Gor V (2000) Great expectations? The dubious financial legacy of quality audits. Br J Manag 11(1):31–45

Aslı Selvi is a PhD student under the Council of Higher Education 100/2000 Boursaryship Programme in "Digital Transformation in Finance and Trade Sectors" at Trakya University. She is pursuing her PhD in the Accounting and Finance Department. She has a master's degree from the Econometrics Program at Dokuz Eylul University. Her bachelor's degree is from Çanakkale Onsekiz Mart University, Department of Econometrics. She works as a project staff at Hermes Consulting Company in Trakya Technopark.

Kıymet Tunca Çalıyurt graduated from Beşiktaş Atatürk Lisesi (1986) in İstanbul. Her bachelor's degree is from the Business Administration Department, Marmara University (1990) in Istanbul; her Master's (1994) and Ph.D. (2002) degrees are from the Accounting and Finance Program at the Social Graduate School, Marmara University; her postdoctoral study is from London Metropolitan University (2004).

Prof Çalıyurt worked as an auditor at Horwath Auditing Company, a manager at McDonald's, and finance staff at Singapore Airlines. After vast experience in the private sector, she started working in academia. Her research interests are accounting, auditing, fraud examination, social responsibility, corporate governance, finance, business ethics, and sustainability, particularly in auditing, aviation management, NGOs, and the fast sector.

Prof Çalıyurt was a visiting researcher at the John Molson School of Business in 2001 summer (International Center for Aviation Management), Massachusetts University Amherst Business School in the 2014 summer, the 2018 summer, 2024 summer (Accounting and Information System), University of National World and Economy in the 2019 summer (Faculty of Accounting). She founded the International Group on Governance, Fraud, Ethics, and Social Responsibility (IGonGFE&SR) and the International Women and Business Group (IWBG) in 2009; both groups have organized annual conferences and published international book series.

Prof Çalıyurt performs as

- Book series editor for Springer (indexed in Scopus), titled "Accounting, Finance, Sustainability, Governance & Fraud: Theory and Application" [1], and for Routledge, titled "Women and Sustainable Business" [2].
- Invited speaker at the Cambridge Economic Crime Symposium and workshop organizer on the role of external auditors in fraud prevention.
- Editorial board member in the Journal of Financial Crime (Web of Science—ESCI), International Journal of Law and Management (Web of Science—ESCI), Journal of Money Laundering Control (Web of Science—ESCI), International Journal of Climate Change Strategies and Management (Web of Science—SSCI).

She is a founding partner of Hermes Consulting, located in Trakya University Technopark. The company provides training and services in company fraud examinations, sustainability, auditing, accountability, and academic publishing.

Part II
Numerical Approach to CSR, Sustainability and Accountability

Chapter 2
Exploring the Association Between Mass Shootings and Police Violence in the United States

Hilal Özen, Tuğba Altıntaş, and Eyüp Çetin

Abstract This study delves into the distinctive attention given to mass shootings and police violence, particularly in the United States. Analyzing cross-sectional 2022 data, we utilize canonical correlation and multiple correspondence analyses to unveil the relationship between these phenomena. A robust canonical correlation of 0.87 ($p < 0.05$) underscores a significant relationship, with the number of people killed in mass shootings influencing police violence by 82.7%, and vice versa at 85.7%. Multiple correspondence analysis further supports this link, revealing synchronous trends in variables such as the number of people killed, incidents, and gun sales in mass shootings with the number of people killed by police in police violence. Our study is the first in the literature to reveal a significant correlation between mass shootings and police violence in the United States.

H. Özen (✉)
Department of Production Management and Marketing, Faculty of Economics and Administrative Sciences, Trakya University, Edirne, Türkiye
e-mail: hilalozen@trakya.edu.tr

T. Altıntaş
Department of Health Management, Faculty of Health Sciences, Üsküdar University, İstanbul, Türkiye
e-mail: tugba.altintas@uskudar.edu.tr

E. Çetin
Department of Quantitative Methods, School of Business, Istanbul University, İstanbul, Türkiye

Laboratory for Industrial and Applied Mathematics, York University, Toronto, ON, Canada

New York Business Global, College Park, MD, USA
e-mail: eyup.cetin@istanbul.edu.tr; eyupc@yorku.ca

K. T. Çalıyurt (ed.), *New Approaches to CSR, Sustainability and Accountability, Volume VI*, Accounting, Finance, Sustainability, Governance & Fraud: Theory and Application, https://doi.org/10.1007/978-981-96-5937-1_2

Our findings: The correlation between mass shootings and police violence does not imply causation. The correlation alone is not enough to prove causality. Further research is needed to establish causality.

Keywords Mass shooting · Police violence · Public health · Canonical correlation analysis · Multiple correspondence analysis

2.1 Introduction

Mass shootings have unfortunately become a common method of violence worldwide. However, when we examine the frequency, it is evident that these incidents are much more prevalent in the United States compared to the rest of the world. Mass shootings pose a serious issue for the United States, resulting in numerous tragic deaths and injuries every year. To better understand the problem, it is essential to define the term.

Krouse and Richardson (2015) provided a definition of mass shootings as "multiple homicide incident in which four or more victims are murdered with firearms, all within a single event and in one or more locations in close proximity."

Since the early 1900s, there has been a documented list of mass shootings occurring in the United States. These incidents have taken place in various public locations, including schools, colleges, universities, restaurants, malls, offices, and festivals (Wikipedia 2022). However, upon examining the "Violence Project" database by Peterson and Densley (2022), it becomes evident that in the last 60 years, mass shootings have predominantly occurred in workplaces and retail settings. The statistics related to fatalities and injuries are staggering. According to the Mother Jones Mass Shooting Database, between 1982 and October 2022, there have been 145 recorded cases, resulting in a total of 1826 victims (Mother Jones 2022).

The importance of mass shooting incidents has gained attention across various academic disciplines. In the last decade, a significant number of articles have focused primarily on mass shootings from different perspectives (Lemieux 2014; Newman and Hartman 2017; Chau 2018; Wallace 2015; Luca et al. 2020; Pearce 2020; Frey and Kirk 2021; Kim et al. 2021; Miller and Bryant 2022; Newsome et al. 2022).

Mass shootings have been the subject of investigation from various angles, including the relationship between mass shootings and gun sales (Wallace 2015; Chau 2018; Callcut et al. 2019; Liu and Wiebe 2019; Osuna 2019; Porfiri et al. 2020; Miller and Bryant 2022; Newsome et al. 2022), the mental health of the shooters, state gun policies (Porfiri et al. 2020), and the influence of traditional and digital media.

In theory, the relationship between mass shootings and police violence is complex and multifaceted. As common factors, both police violence and mass shootings involve the use of lethal force and effected people. Thus, the frequencies and the number of injured and killed people, as well as gun culture, gun violence rate, and gun law strength should be considered in both phenomena for exploring the association. Societal factors such as gun availability, social inequality, and mental health issues may contribute to both phenomena. In police violence, police officers are typically the perpetrators, while mass shootings involve civilians. Hence, the number of police officers and police rate along with the number of people killed by police should be key indicators. Regarding targets, police violence primarily affects individuals involved in law enforcement encounters when mass shootings target random crowds, schools, workplaces, or specific groups. As a result, the numbers of incidents and affected people should be taken into account as factors.

As theoretical potential links between mass shootings and police violence, first, in the dimension of gun culture, it is clear that the prevalence of firearms in the United States may contribute to both police shootings and mass shootings. Also, in a recent study, Geller et al. (2021) highlighted the relationship between domestic violence and mass shootings. Most mass shootings are related to domestic violence and involve other forms of violence, not just public shootings. Consequently, it may be said that exposure to violence, including police violence, may desensitize individuals, potentially contributing to mass shootings. Therefore, the rate of gun violence should be regarded as a factor to be examined in both contexts. Besides, escalation of force by police can lead to violent confrontations, which may influence civilian behavior. In addition, perceptions of injustice due to police violence can fuel anger and resentment, potentially leading to mass shootings. Moreover, Silva (2021) quantitatively compared mass shootings involving police victimized by firearms against all other non-police victimized by firearms. It may be claimed that such historical experiences by police may contribute to potential police violence. Furthermore, in Swedish research by Karlsson (2005), all police officers indicated having vivid recollections of the mass shooting episode, with visual and auditory reminiscences being the most enduring, despite a decrease noted 9 years post-incident. Such past recollections may contribute to potential police violence.

Instead of singular occurrences, adopting a comprehensive strategy at the state level in the United States, this study offers a fresh perspective on mass shootings by examining their relationship with police violence. Surprisingly, the existing literature lacks any articles that explore the connection between mass shootings and police violence. In this study, we consider a range of variables related to mass shootings, including the number of mass shooting incidents, the number of people killed, the number of people injured, gun law strength, gun violence rates, and the number of gun sales. In contrast, when analyzing police violence, we consider variables such as the number of people killed per million by police, the police rate (the ratio of the number of police officers to the state's population per 100,000), and the number of people killed due to police violence. These variables are collected for all 51 states in the United States. Since there was missing data for Hawaii and Washington, DC, those states were eliminated from the data. Apart from causality, we hypothesize that there may be a significant relationship between mass shooting incidents occurring across all states and police violence. We employ cross-sectional data for all variables obtained for the year 2022 which is available as completed year for all variables.

2.2 Related Literature

There are many academic articles regarding mass shooting which examine the flash topic from various perspectives. As mentioned before, some of them found correlation between gun sales and some found relation with the shooters' mental situation. In some articles, the gun policies of the states were examined. The starting point of this paper is to find the relationship between police violence and mass shooting. Before defining the variables that make up those two main factors, we will try to summarize the studies made about mass shooting in the last years.

People's histories and background may play an important role in mass shooting cases, and in some articles, the findings suggest that there can be a correlation between them. Geller et al. (2021) tried to investigate the role of domestic violence (DV) on mass shooting and found that most of the mass shooting incidents were related with DV. In their up-to-date article, Lockwood et al. (2023) also argued that according to the studies of US Secret Service show that most of the mass attackers had a DV background. So, DV is also a related factor with mass shooting. Lankford et al. (2019) came with the question if the mass shooting incidents could be preventable with various signals and they found that offenders were reported to law enforcement more often when compared to other active shooters. On the other hand, Yelderman et al. (2019) tried to understand if the mental health history of mass shooters and their access to firearms are related with the severity of the mass shooting case. As the result of the conducted study, the relations were supported.

Gun law policy actions are also vital in mass shooting for the need of preventing the cases. This leads to research about the policies and regulations. Luca et al. (2020) argued that mass shooting incidents have some impact on gun policy, and one of the remarkable findings show that after a mass shooting incident happens in the United States, there is always an increase in the number of firearm bills introduced. Webster et al. (2020) focused on the gun policies like handgun purchaser licensing laws, assault weapons bans, bans of large-capacity magazines, and deregulation of civilian concealed carry of firearms. Results show that large-capacity magazines bans and handgun purchaser licensing laws help to reduce fatal mass shooting incidents. Post et al. (2021) investigated if the Federal Assault Weapons Ban (FAWB) in the United States was effective in reducing the mass shooting cases. They found that during the FAWB was in place, there was a reduction in the number of mass shooting cases. Koper (2020) examined the regulations for weapons and argues that there is growing evidence which suggests the state-level restrictions reduce mass shooting cases. Siegel et al. (2020) analyzed the mass shooting data in the United States from 1976 to 2018 and found the states that require permission for purchasing firearms were related with 60% lower mass shooting cases. There are also some different findings about the results of the bans and regulations regarding firearms. For example, Greene-Colozzi and Silva (2020) examined the mass shootings from 1966 to 2018 and found that if there are bans for assault weapons, the illegal obtainment of those weapons increases. So, sometimes the restrictions and regulations increase the illegal situations.

When the ownership and sales of guns were checked, there are also several studies that examined the relationship. For example, Fridel (2021) found that the states in which the gun ownership levels are high, mass shooting cases occur disproportionately. Looking at the issue from another perspective, some studies have investigated the patterns of gun sales following mass shooting incidents. Iwama and McDevitt (2021), for instance, conducted a study using statewide data from Massachusetts spanning from 2006 to 2016. They discovered that after each mass shooting event, there was a significant increase in handgun sales in the following week. Similarly, Callcut et al. (2019) conducted research into mass shooting incidents, using California Department of Justice gun sales reports. Their findings revealed a significant increase in gun sales after each mass shooting incident. Even as California authorities enacted stricter gun laws in response to these incidents, gun sales continued to rise.

2.3 Materials and Methods

This section is structured into the respective categories of Data Sources and Methodologies.

2.3.1 Data Sources

In this research, two sets of variables have been formulated, one pertaining to mass shootings (MS) and the other related to police violence (PV). We give and explain our data sources for both mass shooting and police violence as follows.

2.3.1.1 Mass Shooting

Number of Mass Shooting Incidents, Number of People Killed, and Number of People Injured

These three variables were sourced from the Gun Violence Archive website (gunviolencearchive.org). This website maintains a comprehensive archive of mass shooting incidents across the United States, organized by state. The report for mass shootings in 2022 summarizes these incidents by their Incident ID, Incident Date, State, City or County, Address details, and includes information on the number of people killed and the number of people injured. Consequently, our data on the number of mass shooting incidents, as well as the number of people killed and injured, is directly derived from this database.

Gun Law Strength and Gun Violence Rates

The assessment of mass shootings in all states is greatly influenced by the strength of their gun laws and the prevalence of gun violence. To obtain this data, we sourced information from Everytown Research's "Everytown Gun Law Rankings" (everytownresearch.org). Everytown Research calculates the strength of gun laws by conducting a head-to-head analysis of the 50 most crucial gun safety policies, with the highest possible score being 100. Additionally, the handgun violence rate is determined by gun deaths per 100,000 residents, as per our data source.

Number of Gun Sales

The state-level gun sales data were sourced from research conducted by Nass and Barton (2023) available on thetrace.org. This research comprehensively analyzed data for both nationwide and state-level gun sales on an annual basis. The database includes records from 2001 to 2023, and we specifically examined gun sales trends for the year 2022.

Surprisingly, the impact of COVID-19 led to a significant surge in gun sales in the United States in 2020. Many have referred to that year as a "record-breaking" year for gun sales.

2.3.1.2 Police Violence

Number of People Killed Per Million

We obtained data on the number of people killed due to police violence in the United States from Mapping Police Violence (mappingpoliceviolence.org). This website provides users with the ability to filter data from 2013 to 2023. For our study, we focused on the year 2022.

Number of People Killed by Police

According to The Washington Post (2023), it has been tracking fatal police shootings in the United States since 2015 and reports that the police have been responsible for more than a thousand deaths each year. This number has shown an annual increase, and we specifically used data from the year 2022 for our analysis.

Number of Police Officers and Population

In the police violence data set, the number of police officers and the population for each state are crucial variables. The population data for U.S. states were sourced from The US Census Bureau database for the year 2022 (US Census Bureau 2023).

Furthermore, the number of police officers was obtained from the US Bureau of Labor Statistics, also for the year 2022 (US Bureau of Labor Statistics 2023).

In this study, as we set before, two sets of variables have been created, one is for mass shooting (MS) and the other one is for police violence (PV). The mass shooting variable set consists of the following variables: the number of mass shooting incidents, the number of people killed, the number of people injured, gun law strength, gun violence rate, and gun sales. The police violence variable set includes the following variables: the number of people killed by police per million, the number of people killed by police, and the number of police per 100,000 people. The study used data of the year 2022 for 49 states of the United States. The descriptive statistics for the variables comprising both sets are presented in Table 2.1.

2.3.2 *Methodologies*

The main objective of the study is to explore the association between mass shooting and police violence. To achieve this goal, canonical correlation analysis is initially used. The Canonical Correlation Analysis (CCA) is a statistical technique used to investigate the relationships between two sets of variables. It aims to identify and quantify the associations between two multivariate sets of variables while maximizing the correlation between linear combinations of these sets. In CCA, the method seeks linear combinations of variables, known as canonical variables, in each set to maximize the correlation between them. These canonical variables represent new dimensions that capture the highest possible correlation between the two sets. The canonical loadings and canonical correlations can be examined to understand which variables contributed the most to the observed associations and the strength of these relationships (Hair 2014; Johnson and Wichern 2007).

Table 2.1 Descriptive statistics

Variable	Mean	Standard deviation	Minimum	Maximum
Mass shooting				
MS incident	1267	1396	00	5700
MS killed	1259	1657	00	8400
MS injured	5214	5776	00	25,400
MS gun law strength	3134	2491	300	8450
MS gun violence rate	1513	617	340	2860
MS sales	33,673,688	29,541,802	2,744,100	140,310,100
Police violence				
PV killed per million	616	1143	71	8300
PV killed by police	2149	2505	100	12,800
PV police officer per 100k	20,171	9377	11,456	80,951

As a complementary non-parametric approach, in addition to canonical correlation analysis, which is a parametric method, the variables are transformed into categorical form, and multiple correspondence analysis is used to investigate other relationships between the categories. Each variable has been divided into three groups: "low," "moderate," and "high" based on the values they take. This procedure is required to employ Multiple Correspondence Analysis (MCA) which is a statistical method commonly used to explore and analyze the relationships between categorical variables. It is an extension of Correspondence Analysis (CA) and is particularly valuable when dealing with multi-way contingency tables. In MCA, the method constructs a low-dimensional graphical representation of the categorical variables and their categories in a multi-dimensional space. Each category is mapped to a point in this space, and the distances and angles between the points reflect the strength and nature of the associations between the categories (Le Roux and Rouanet 2010). Consequently, this later analysis facilitates the investigation of the relationship between MS and PV through a non-parametric approach.

2.4 Results and Discussions

The results of our analyses can be examined in two distinct categories.

2.4.1 Findings of Canonical Correlation Analysis (CCA)

An exploration of the multivariate relationship using CCA reveals three canonical correlations, with one significant ($p < 0.05$) canonical function for the relationship between mass shooting and police violence (Table 2.2).

The first canonical function demonstrates a strong relationship between the two sets of variables with a coefficient of 0.870. The corresponding Wilks statistic of 0.170 indicates a significant relationship ($p < 0.05$), too. This study focuses only on the exploration of canonical loadings and cross-loading for the first canonical function which is significant only.

Canonical loadings can be used for exploring the multivariate correlations between variables and their corresponding canonical variates. As shown in Table 2.3,

Table 2.2 Canonical correlations

	Correlation	Eigenvalue	Wilks Statistic	F	Num D.F	Denom D.F.	Sig.
1	870	3119	170	5496	18,000	113,622	000
2	492	320	701	1597	10,000	82,000	122
3	275	082	924	859	4000	42,000	497

the most effective variable in Set 1 (MS) is the number of people killed with the greatest canonical loading 95.1%; it is, respectively, followed by gun sales, the number of incidents, the number of people injured, gun law strength, and gun violence rate. The most effective variable in Set 2 (PV) is the number of people killed by police with the greatest canonical loading 98.5%; it is, respectively, followed by the number of police officers per 100k, and the number of people killed per million. It is observed that the number of people killed of MS is the most effective variable on Set 2 (PV) with 82.7% cross-loading. It is, respectively, followed by gun sales, the number of incidents, the number of people injured, gun law strength, and gun violence rate. The number of people killed by police of PV is the most effective variable on Set1 (MS) with 85.7% cross-loading. It is, respectively, followed by the number of police officers per 100k and the number of people killed per million.

Also, the proportion of the explained variances is given in Fig. 2.1.

The variance of MS is explained by its variables with a proportion of 54%. The variance of MS is explained by PV variables with a proportion of 41%. The variance of PV is explained by its variables with a proportion of 34%. The variance of PV is explained by MS variables with a proportion of 26%.

2.4.2 *Findings of Multiple Correspondence Analysis (MCA)*

Table 2.4 provides a summary of the model derived from multiple correspondence analysis, indicating the significance of each dimension.

The table displays Cronbach's alpha values for Dimension 1 (0.821) and Dimension 2 (0.686). These values, which fall within the range of 0 to 1, reflect the degree of internal consistency among the categories within each respective dimension, with higher values signifying a more robust level of internal consistency. The eigenvalues provide insights into the extent of variance accounted for by each dimension. Specifically, Dimension 1 exhibits an eigenvalue of 3.549, when Dimension 2 is associated with an eigenvalue of 2.499. Larger eigenvalues signify that the respective dimension explains a greater proportion of the variance within

Table 2.3 Canonical loadings and canonical cross-loadings

Canonical variate	Variable	Canonical loading	Cross-loading
MS	Incident	866	754
	Killed	951	827
	Injured	833	725
	Gun law strength	143	124
	Gun violence rate	056	049
	Sales	918	798
PV	Killed per million	−172	−149
	Killed by police	985	857
	Police officer per 100k	181	157

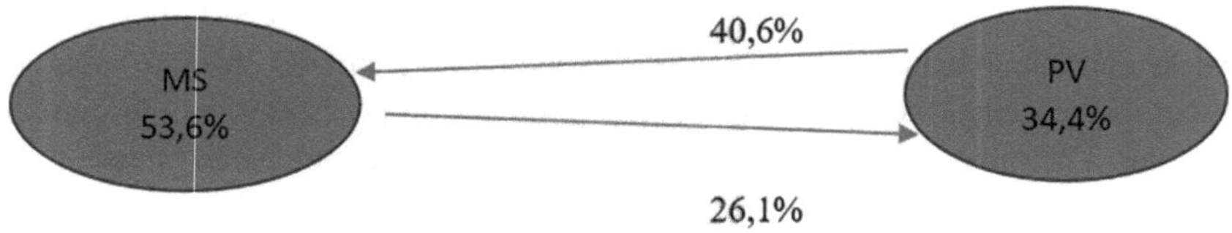

Fig. 2.1 Proportion of the explained variances

Table 2.4 Multiple correspondence analysis model summary

Model summary				
		Variance accounted for		
Dimension	Cronbach's alpha	Total (Eigenvalue)	Inertia	% of variance
1	821	3549	444	44,363
2	686	2499	312	31,238
Total		6048	756	
Mean	765[a]	3024	378	37,800

[a] Mean Cronbach's alpha is based on the mean eigenvalue

the data set. Inertia serves as a measure of the overall variance accounted for by each dimension. As indicated in the table, Dimension 1 explains 44.4% of the total inertia, while Dimension 2 accounts for 31.2%. These percentages provide valuable insights into the extent to which each dimension captures and explains the information within the dataset. As a noteworthy discovery, Dimension 1 accounts for 44.4% of the variance, while Dimension 2 explains 31.2%. Together, these two dimensions collectively elucidate 75.6% of the overall variance, as demonstrated in the "Total" row. The values in the table indicate the suitability of the method.

In the following Fig. 2.2, the positions of the categories in a two-dimensional space are displayed.

The category marked with 1 in the figure indicates a low value for that variable, the category marked with 2 represents a moderate value, and the category marked with 3 signifies a high value.

Usually, there is a noticeable consistency in the levels of the categories. For instance, concerning the left side of the first dimension, when the MS variables "killed," "incident," and "sales" are low, the PV variable "killed by police" is also low. On the right side, when the MS variables "killed," "incident," and "sales" are high, the PV variable "killed by police" is also high. It can be noted that the variables "killed," "incident," and "sales" in mass shootings exhibit a parallel trend and correlation with the "killed by police" variable in police violence.

The "low" category of the police officers per 100,000 variables is nearly equidistant from all categories of the other variables. In both dimensions, the "moderate" level categories of the variables "gun violence rate," "incident," "killed," and "sales" from the MS variables, as well as "killed by police" and "sales" from the PV variables, are closely situated around the origin.

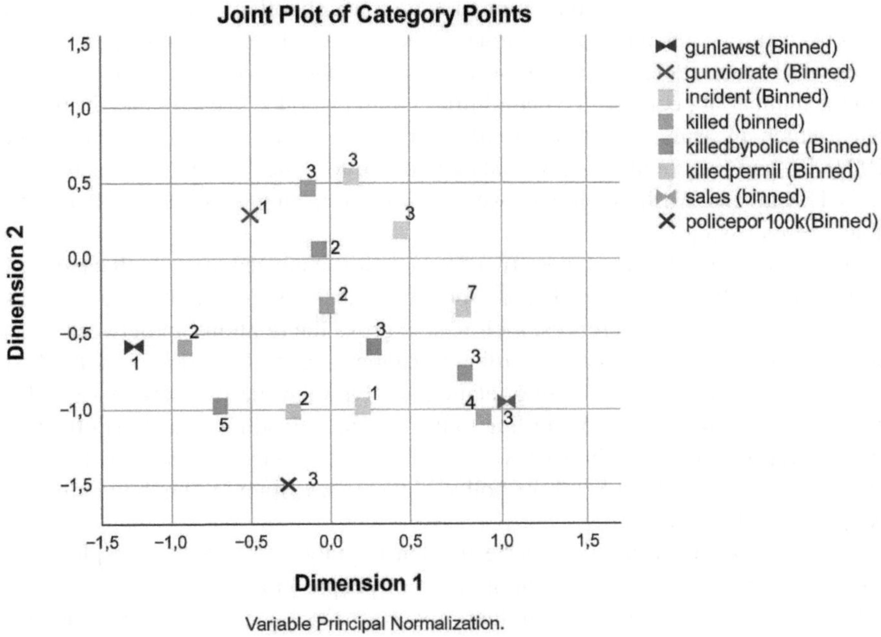

Fig. 2.2 Positions of the categories in a two-dimensional space

The most remarkable pattern in the figure is that the values of "gun violence rate" and "killed per million" are low in the PV category when "gun law strength" is high, as observed in the lower part of the figure in relation to the second dimension. Likewise, when "gun law strength" is low, there is a notable pattern where "gun violence rate" and "killed per million" in the PV category tend to be high, as observed above the origin.

2.5 Conclusion

The escalating frequency of mass shootings in the United States remains a chronic issue, as evidenced by sources (Statista 2023; everytownresearch, 2023). This topic warrants comprehensive investigation from diverse perspectives. While the impact of certain variables in triggering mass shootings is a trending subject, an equally intriguing area lies in examining these variables'interrelationships. This paper considers police violence has a significant relationship with mass shootings. The Washington Post (2023) has documented annual and monthly fatalities resulting from police encounters. Thus, there is an assumption that a correlation might exist between mass shootings and incidents of police violence.

The variables pertaining to mass shootings encompassed several metrics: the count of mass shooting incidents, the number of people killed resulting from mass shootings, the number of people injured from mass shootings, the strength of gun laws in each state, the incidence of gun violence in each state, and the volume of gun sales within each state. In contrast, the analysis of police violence involved examining the number of people killed per million caused by police in each state, the police rate (which measures the ratio of police officers to the state's population per 100,000), and the count of individuals killed due to police violence in each state. Only Hawaii and Washington, DC, were eliminated from the data, since their variables were missing.

The relationship between these two data sets was analyzed using both canonical correlation analysis and multiple correspondence analysis. Although these analyses have methodological differences, their results prove the relationship between mass shooting and police violence. According to canonical correlation analysis, the first canonical function shows a strong relationship between the two sets of variables whose coefficient is 0.870. According to the canonical loadings, the most effective variable for mass shooting in Set 1 was found to be the number of people killed. The most effective variable in Set 2 for police violence was found to be the number of people killed by police.

On the other hand, multiple correspondence analysis was held to explore the relationship between two sets. The number of people killed, the number of incidents, and gun sales variables from mass shooting move in the same direction with the number of people killed by police from police violence, which also reveal the relationship between them.

This study has delved into the correlation between mass shootings and police violence, employing multiple analytical methods on data gathered from the year 2022. The correlation between these phenomena does not necessarily indicate causation in either case. The findings suggest a notable trend: as instances of police violence increase, so do the occurrences of mass shootings. To address this concerning association, proactive measures targeting the reduction of gun sales are imperative. The observed link between increased gun sales and the rise in both mass shootings and fatalities resulting from police encounters underscores the urgency for strategic interventions.

Furthermore, future research endeavors should extend beyond the variables explored in this study. Investigating additional factors within these data sets and analyzing data from other years could provide a more comprehensive understanding of the nuanced dynamics between mass shootings and police violence. Such investigations will contribute significantly to devising effective strategies and policies aimed at curbing these alarming trends.

Appendix: Data Used for the Analyses

	Mass shooting						Police violence					
State	# Incident	# Killed	# Injured	Gun law strength	Gun violence rate	Sales	Killed per mn	# Killed by police	# Police officers	Population	# Police off/ population	Per 100k
Mississippi	14	15	55	3	28.6	249,183	5.07	13	7520	2,940,820.00	0.002557	255.711
Wyoming	**0**	**0**	**0**	6	25.9	66,755	17.34	10	1230	581,813.00	0.002114	211.4081
Louisiana	28	14	135	20	26.3	320,645	4.51	19	13,080	4,576,882.00	0.002858	285.7841
Alaska	**0**	**0**	**0**	9	23.5	84,868	10.91	5	1310	731,721.00	0.00179	179.03
Alabama	16	11	67	18	23.6	501,499	4.98	20	12,570	5,056,005.00	0.002486	248.6153
Missouri	14	13	60	9	23.9	528,187	5.52	29	12,300	6,180,523.00	0.00199	199.0123
New Mexico	4	2	16	37.5	22.7	164,052	15.11	30	4610	2,113,550.00	0.002181	218.1164
Arkansas	5	8	35	5	22.6	200,285	4.98	16	5200	3,042,231.00	0.001709	170.9272
Montana	**0**	**0**	**0**	5	20.9	130,412	6.46	7	1940	1,125,473.00	0.001724	172.372
South Carolina	20	15	93	18	22	270,118	5.08	21	11,640	5,259,222.00	0.002213	221.3255
Tennessee	17	14	71	16.5	21.3	638,652	4.78	30	12,650	7,059,618.00	0.001792	179.1882
Oklahoma	8	21	19	7.5	20.7	332,377	7.83	28	8900	4,018,931.00	0.002215	221.4519
Kentucky	9	7	39	9	20.1	296,576	4.22	16	6460	4,519,315.00	0.001429	142.942
West Virginia	**0**	**0**	**0**	20	18.1	166,691	7.25	12	3190	1,773,866.00	0.001798	179.8332
Georgia	29	30	106	11.5	17.7	436,261	5.79	49	21,140	10,852,484.00	0.001948	194.7941
Idaho	**0**	**0**	**0**	5	17.6	171,212	4.35	7	3160	1,981,332.00	0.001595	159.4887
Nevada	5	2	35	34	17	119,846	5.8	16	5620	3,201,212.00	0.001756	175.5585
Indiana	15	17	56	22	17.3	522,143	2.65	17	12,470	6,831,167.00	0.001825	182.5457
Arizona	13	13	67	8.5	16.7	424,364	8.11	52	11,850	7,408,017.00	0.0016	159.9618
Kansas	3	3	12	9.5	16.9	179,368	4.77	12	5920	2,928,419.00	0.002022	202.1569

(continued)

	Mass shooting						Police violence					
State	# Incident	# Killed	# Injured	Gun law strength	Gun violence rate	Sales	Killed per mn	# Killed by police	# Police officers	Population	# Police off/ population	Per 100k
North Carolina	21	17	87	31	16	223,166	3.64	32	20,480	10,678,831.00	0.001918	191.7813
Colorado	13	23	58	58	15.4	484,191	6.75	40	10,140	5,855,078.00	0.001732	173.183
Ohio	19	19	75	18	15.2	589,038	3.47	38	24,150	11,767,059.00	0.002052	205.234
Michigan	25	19	98	25.5	14.6	547,201	2.08	20	16,350	10,032,720.00	0.00163	162.9668
Texas	50	84	199	13.5	14.2	1,403,101	4.36	118	59,290	29,947,238.00	0.00198	197.9815
Florida	31	21	126	33.5	13.7	1,239,800	3.58	62	47,000	22,040,323.00	0.002132	213.2455
Oregon	6	5	23	54	13	449,938	5.43	23	4880	4,259,743.00	0.001146	114.5609
Illinois	57	50	254	74	14.1	463,904	2.19	24	29,290	12,518,144.00	0.00234	233.9804
Pennsylvania	34	21	145	39.5	13.6	903,000	2.08	23	25,210	12,916,089.00	0.001952	195.1829
Delaware	1	0	4	50	14.4	55,329	2.02	2	1860	1,012,615.00	0.001837	183.6828
Virginia	20	36	75	48	13.4	562,876	3.36	24	17,190	8,662,151.00	0.001984	198.4496
South Dakota	**0**	**0**	**0**	5.5	13.6	79,945	7.89	7	1830	905,404.00	0.002021	202.1197
Maryland	19	29	52	71.5	13.5	212,420	1.94	11	10,280	6,157,731.00	0.001669	166.9446
Utah	**0**	**0**	**0**	11	13.6	143,706	3.97	11	5200	3,426,431.00	0.001518	151.7614
North Dakota	**0**	**0**	**0**	11.5	13.8	66,574	3.85	3	1710	770,221.00	0.00222	222.0142
Wisconsin	16	16	62	27.5	12.2	440,492	2.38	12	11,530	5,903,573.00	0.001953	195.3055
Vermont	**0**	**0**	**0**	30.5	11.6	48,688	3.11	2	870	647,378.00	0.001344	134.3883
Maine	**0**	**0**	**0**	20.5	10.4	114,829	4.40	5	2080	1,381,441.00	0.001506	150.5674
Washington	11	4	49	59.5	10.9	473,349	5.45	39	8960	7,765,004.00	0.001154	115.3895
Iowa	2	3	13	14.5	11.2	82,903	1.88	6	5090	3,196,911.00	0.001592	159.2162
Nebraska	5	6	17	32	10.1	27,441	2.04	4	3320	1,966,441.00	0.001688	168.8329
New Hampshire	**0**	**0**	**0**	9	8.9	128,754	2.18	4	2900	1,405,243.00	0.002064	206.37
Minnesota	12	10	43	37.5	8.9	324,510	1.58	9	9020	5,702,253.00	0.001582	158.1831
California	49	60	186	84.5	8.5	843,351	3.69	128	70,090	38,959,247.00	0.001799	179.9059

	Mass shooting						Police violence					
State	# Incident	# Killed	# Injured	Gun law strength	Gun violence rate	Sales	Killed per mn	# Killed by police	# Police officers	Population	# Police off/ population	Per 100k
Connecticut	3	1	11	75.5	6	99,384	0.83	1	6410	3,609,924.00	0.001776	177.5661
New York	31	23	129	78	5.3	364,297	1.34	19	50,600	19,415,392.00	0.002606	260.618
Rhode Island	**0**	**0**	**0**	52	5.1	31,223	0.91	1	1760	1,094,843.00	0.001608	160.7536
New Jersey	9	3	40	71	5	171,895	0.97	9	20,510	9,242,109.00	0.002219	221.919
Massachusetts	3	0	14	77	3.7	121,308	0.71	5	17,450	6,962,372.00	0.002506	250.633
Hawaii	**0**	**0**	**0**	79.5	3.4	NR	1.37	2	2360	1,433,336.00	0.001647	164.6509
District of Columbia (Washington, DC)	9	7	33				7.25	3	5320	657,185	0.008095	809.5133

References

Callcut RA, Robles AM, Kornblith LZ, Plevin RE, Mell MW (2019) Effect of mass shootings on gun sales—a 20-year perspective. J Trauma Acute Care Surg 87(3):531–540. https://doi.org/10.1097/TA.0000000000002399

Chau, E. (2018), "The Effects of Mass Shootings on Gun Sales", College of the Holy Cross, Department of Economics, *Faculty Research Series*, Paper No. 1804*

Everytown Research. Gun safety policies save lives. https://everytownresearch.org/rankings/. Accessed 14 Oct 2022

Everytown Research. Mass Shootings in the United States. https://everytownresearch.org/mass-shootings-in-america/. Accessed 20 Nov 2023

Frey A, Kirk DS (2021) The impact of mass shootings on attitudes toward gun restrictions. Socius: Sociological Research for a Dynamic World 7:1–9. https://doi.org/10.1177/23780231211054636

Fridel EE (2021) Comparing the impact of household gun ownership and concealed carry legislation on the frequency of mass shootings and Firearms Homicide. Justice Q 38(5):892–915. https://doi.org/10.1080/07418825.2020.1789693

Geller LB, Booty M, Crifasi CK (2021) The role of domestic violence in fatal mass shootings in the United States, 2014–2019. Injury Epidemiology 8(38)

Greene-Collozi EA, Silva JR (2020) Contextualizing firearms in mass shooting incidents: a study of guns, regulations, and outcomes. Justice Q 39(4):697–721. https://doi.org/10.1080/07418825.2020.1818805

Gun Violence Archive. "Mass shootings in 2020". https://www.gunviolencearchive.org/reports/mass-shooting?year=2020. Accessed 12 Oct 2022

Hair JF (ed) (2014) Multivariate data analysis, 7th edn. Pearson

Iwama J, McDevitt J (2021) Rising gun sales in the wake of mass shootings and gun legislation. J Prim Prev 42:27–42. https://doi.org/10.1007/s10935-021-00622-7

Johnson RA, Wichern DW (2007) Applied multivariate statistical analysis, 6th edn. Pearson Prentice Hall

Karlsson I (2005) Memories of traumatic events among Swedish police officers, Ph.D. Dissertation, Stockholm University, Faculty of Social Sciences, Department of Psychology

Kim C, Capellan JA, Adler A (2021) Exploring the empirical literature on mass shooting: A mixed-method systematic review of peer-reviewed journal articles. Aggress Violent Behav 58. https://doi.org/10.1016/j.avb.2021.101584

Koper CS (2020) Assessing the potential to reduce deaths and injuries from mass shootings through restrictions on assault weapons and other high-capacity semiautomatic firearms. Criminology & Public Policy 19(1) Special Issue: Countering Mass Violence in the United States:147–170. https://doi.org/10.1111/1745-9133.12485

Krouse WJ, Richardson DJ (2015) "Mass murder with firearms: Incidents and victims", 1999–2013. Congressional Research Service, Washington, DC

Lankford A, Adkins KG, Madfis E (2019) Are the deadliest mass shootings preventable? An assessment of leakage, information reported to law enforcement, and firearms acquisition prior to attacks in the United States. J Contemp Crim Justice 35(3). https://doi.org/10.1177/1043986219840231

Le Roux B, Rouanet H (2010) Multiple correspondence analysis. SAGE Publications, Inc. https://doi.org/10.4135/9781412993906

Lemieux F (2014) Effect of gun culture and firearm laws on gun violence and mass shootings in the United States: a multi-level quantitative analysis. Int J Crim Justice Sci 9(1):74–93

Liu G, Wiebe DJ (2019) A time-series analysis of firearm purchasing after mass shooting events in the United States. JAMA Netw Open 2(4):e191736. https://doi.org/10.1001/jamanetworkopen.2019.1736

Lockwood KL, Rowe C, Sager E (2023) Policy recommendations to address the nexus of domestic violence and gun violence. North Carolina Medical Journal 84(4). https://doi.org/10.18043/001c.81269

Luca M, Malhotra D, Poliquin C (2020) The impact of mass shootings on gun policy. J Public Econ 181. https://doi.org/10.1016/j.jpubeco.2019.104083

Mapping Police Violence. mappingpoliceviolence.org, Accessed 10 Sept 2023

Miller CE, Bryant A (2022) Evaluation of the association between the flow of firearms and mass shooting deaths. J Consum Aff 56:925–937. https://doi.org/10.1111/joca.12409

Mother Jones, Mass Shooting Database. https://www.motherjones.com/politics/2012/12/mass-shootings-mother-jones-full-data/. Accessed 14 Nov 2022

Nass D, Barton C (2023) How many guns did americans buy last month? https://www.thetrace.org/2020/08/gun-sales-estimates/#:~:text=Americans%20bought%20an%20estimated%201.38,an%20analysis%20of%20FBI%20data.&text=That's%20about%20the%20same%20as,-like%20mass%20shootings%20and%20elections. Accessed 5 May 2023

Newman BJ, Hartman TK (2017) Mass shootings and public support for gun control. Br J Polit Sci 49:1527–1553. https://doi.org/10.1017/S0007123417000333

Newsome K, Sen-Crowe B, Autrey C, Alfaro S, Levy M, Bilski T, Ibrahim J, Elkbuli A (2022) A closer look at the rising epidemic of mass shootings in the United States and its association with gun legislation, laws, and sales. J Surg Res:103–113. https://doi.org/10.1016/j.jss.2022.07.012

Osuna MJR (2019) Mass shootings and gun sales: a study on the influence of red and blue power. University of South Florida, Graduate Theses and Dissertations. https://scholarcommons.usf.edu/etd/7914

Pearce JS (2020) Instagram's affordances and the visual construction of solidarity after a mass shooting. Illn Crisis Loss 30(2):1–15. https://doi.org/10.1177/1054137320906405

Peterson J, Densley J (2022) The violence project: how to stop a mass shooting epidemic. https://www.theviolenceproject.org/. Accessed 5 June 2023

Porfiri M, Barak-Ventura R, Marin MR (2020) Self-protection versus fear of stricter firearm regulations: examining the drivers of firearm acquisitions in the aftermath of a mass shooting. Patterns 1. https://doi.org/10.1016/j.patter.2020.100082

Post L, Mason M, Singh LN, Wleklinski NP, Moss CB, Mohammad H, Issa TZ, Akhetuamhen AI, Brandt CA, Welch SB, Oehmke JF (2021) Impact of Firearm surveillance on gun control policy: regression discontinuity analysis. JMIR Public Health Surveillance 7(4):e26042. https://doi.org/10.2196/26042

Siegel M, Goder-Reiser M, Duwe G, Rocque M, Fox JA, Fridel EE (2020) The relation between state gun laws and the incidence and severity of mass public shootings in the United States, 1976–2018. Law Hum Behav 44(5):347–360. https://doi.org/10.1037/lhb0000378

Silva JR (2021) Police victimized by firearms during mass shootings (1966–2019). J Threat Assess Manag 8(1–2):48–64. https://doi.org/10.1037/tam0000161

Statista, Number of Mass Shootings in the United States between 1982 and October 2023. https://www.statista.com/statistics/811487/number-of-mass-shootings-in-the-us/. Accessed 20 Nov 2023

The Washington Post (2023) 1.009 people have been shot and killed by police in the past 12 months. https://www.washingtonpost.com/graphics/investigations/police-shootings-database/. Accessed 25 May 2023

United States Census Bureau. https://www.census.gov/. Accessed 20 May 2023

US Bureau of Labor Statistics (2023) Occupational employment and wage statistics. https://www.bls.gov/oes/current/oes333051.htm. Accessed 20 May 2023

Violence Policy Center. State Firearm death rates, ranked by rate, 2020. https://vpc.org/state-firearm-death-rates-ranked-by-rate-2020/. Accessed 10 Oct 2022

Wallace LN (2015) Responding to violence with guns: mass shooting and gun acquisition. Soc Sci J 52(2):156–167. https://doi.org/10.1016/j.soscij.2015.03.002

Webster DW, McCourt AD, Crifasi CK, Booty MD, Stuart EA (2020) Evidence concerning the regulation of firearmsdesign, sale, and carrying on fatal mass shootings in the United States. Criminol Public Policy 19:171–212. https://doi.org/10.1111/1745-9133.12487

Wikipedia (2022) List of mass shootings in the United States. https://en.wikipedia.org/wiki/List_of_mass_shootings_in_the_United_States

Yelderman LA, Joseph JJ, West MP, Butler E (2019) Mass shootings in the United States: Understanding the importance of mental health and firearm considerations. Psychol Public Policy Law 25(3):212–223. https://doi.org/10.1037/law0000200

Hilal Özen holds an honors degree in Business Administration (B.S.) from Hacettepe University (2004). She earned her M.S. degree in 2007 and her Ph.D. in 2011, both in Marketing, from Istanbul University School of Business. Currently, she serves as a Professor of Marketing at Trakya University. Her current research interests are focused with great enthusiasm on various areas, including digital marketing, e-health, social media marketing, tourism marketing, sustainability, and consumer decision-making styles. She has published numerous articles on these topics in reputable peer-reviewed journals and presented her research at international conferences. Throughout her academic career, she has been actively involved in teaching a wide range of courses, including principles of marketing, marketing management, research methodology, marketing strategies, current issues in marketing, digital marketing, and healthcare marketing. Her teaching experience spans across all levels of business students, including undergraduate, MS, MBA, and executive MBA students. In addition to her academic pursuits, Hilal is a Co-founder of Coreborn Software and Digital Marketing Company, where she provides valuable consultancy in the field of digital marketing to clients.

Tuğba Altıntaş graduated from İstanbul University, Faculty of Business Administration, with a degree in Business Administration in 2002. She went on to earn her master's degree in Applied Statistics from Ondokuz Mayıs University in 2006. In 2012, she completed her doctorate in Numerical Methods at Istanbul University, earning her Ph.D. She is currently Professor of Operations Research and Statistics and Professor of Health Management in the Department of Health Management in the Faculty of Health Sciences at Üsküdar University. She is also a former Vice Dean of the Faculty of Health Sciences and Director of the Center for Data at Üsküdar University.

Eyüp Çetin is a tenured Professor of Operations Research and Statistics in the Department of Quantitative Methods at the School of Business, Istanbul University, and an Affiliate Distinguished Professor at York University, Canada. He is also affiliated with New York Business Global, USA. He has held various managerial positions. He is former and founding Dean of the Faculty of Transportation and Logistics and former Vice Dean of the School of Business at Istanbul University. He received a Ph.D. in quantitative methods from Istanbul University in 2004. He has taught courses in management science, statistics, business analytics, stochastic models, game theory and health analytics to all levels of business, engineering, healthcare management students: undergraduates, MBAs and Ph.D. students. He has been a consultant for numerous companies, and he has taught statistical & decision modelling and business analytics in even global companies such as Vodafone. His current research focuses on business analytics, data science, digital business & health, healthcare management science, mathematical medicine, operations research, mathematical & statistical modelling. Dr. Çetin is also Founder and Editor-in-Chief of prestigious international academic journals such as European Journal of Pure & Applied Mathematics.

Chapter 3
Diagnosing the Importance of Functions of Organizational Culture on the Recruitment Policy and the Performance of an Organization: Key Indicators for Sustainable Corporate Management

Rauf Nişel

Abstract The aim of the research is to determine the importance of issues that deal with the functions of culture in recruitment process and in performance analysis of companies. In our research, issues that deal with functions of organizational culture are considered and are based on Schein's conceptual approach to organizational culture. Research has been carried out at a company level with separate samples of 213 managers and 841 employees of a leading company in electronic industry. Importance of issues was assessed by principal components analysis. Results showed that the most important issue in our empirical model for both managers and employees was communication. Communication was considered as an important vehicle for conveying explicit cultural values to newcomers and employees with the goal to ensure consistent performance and work standards. Transmission of core values through communication enables sustainable corporate management. Teamwork and performance reward expectancy were other issues which had a different importance from the perspective of managers and employees. Teamwork will facilitate their socialization process by mutual relations with other members of the team. Hence, the newcomers can make adjustments in their job and organization to accommodate their knowledge and personalities. Teamwork is essential for sustainability of corporate management by reducing role ambiguity and role conflict in the organization. Organizations must continuously examine their reward systems to find ways to make it more effective in improving employee performance in order to achieve sustainable corporate management.

R. Nişel (✉)
Department of Industrial Engineering, Faculty of Engineering, Piri Reis University, İstanbul, Türkiye
e-mail: rnisel@pirireis.edu.tr

K. T. Çalıyurt (ed.), *New Approaches to CSR, Sustainability and Accountability, Volume VI*, Accounting, Finance, Sustainability, Governance & Fraud: Theory and Application, https://doi.org/10.1007/978-981-96-5937-1_3

Keywords Recruitment · Organizational culture · Performance · Sustainable corporate management

3.1 Introduction

Companies that manage high levels of employee growth focus on the qualification of potential employees' fit with organizational culture. These companies show greater emphasis on managing culture in positive, explicit ways because people understand the impact of culture on their work and organizational goals. In some companies, the recruitment process is haphazard, inconsistent affair resulting in short-term hires or chronically high turnover. Companies that manage high levels of employee growth successfully take coordinated, systematic approach to recruitment. This formalization of the recruitment process deals with the following steps:

- Determine whether new employees fit with the organizational culture before hiring them.
- Being more consistent in the application process.
- Making clearer and more consistent messages that recruiters tell prospective employees.
- Using information about skill gaps of employees as a part of employee screening and recruitment process.

Companies that manage high levels of employee growth and performance focus on the qualification of potential employees' fit with organizational culture. These companies show greater emphasis on managing culture in positive, explicit ways because people understand the impact of culture on their work and organizational goals. More companies are devoting resources such as time and money to managing organizational culture. There is more emphasis on employee's style and fit with the culture.

Organizational culture has a key role of integration of new employees into the organization. Possible recruitment strategies can be derived from issues that deal with functions of organizational culture. Schein suggests that organizational culture form in response to two major challenges that confront every organization.

1. External adaptation and survival
2. Internal integration (Schein, E.H. 1988)

External adaptation and survival has to do with how organization will cope with its constantly changing environment. Other function of the organizational culture serves for the internal integration of the organization. Internal integration concerns the establishment and maintenance of effective working relationship which can be considered as key indicators of sustainable management of an organization. Internal integration involves addressing following issues as common language and conceptual categories, group and team boundaries, power and status, rewards and punishments.

- **Common *language and conceptual categories:*** To identify methods of communication and to develop a shared meaning for important concepts, employees in the organization must come together to establish a communication system and a language by considering frequency and timing of communication, time allocated to communication, and quality of content of the messages in the communication. The quality of content of messages is defined by its ability to effectively convey the intended information to the target audience in a clear, accurate, relevant, and comprehensive manner. High-quality messages are those that minimize misunderstandings and ambiguities, are tailored to the audience's needs and expectations, and provide all necessary information for the recipient to make informed decisions or take desired actions (Keyton, J. 2011).

  Impact of Communication on Recruitment Policy:

1. Attracting the Right Candidates: High-quality content in recruitment messages ensures that job advertisements are clear, informative, and appealing to the target audience. This helps in attracting candidates who are not only qualified but also a good fit for the organization's culture and values. Clear and detailed job descriptions can prevent applications from unqualified candidates and attract those who are genuinely interested and capable (Breaugh, J. A. 2008).
2. Employer Branding: The quality of messages reflects on the employer's brand and reputation in the job market. Organizations that consistently communicate clearly and effectively are perceived as more desirable employers. This positive perception helps attract top talent and can give an organization a competitive edge in the recruitment market (Backhaus, K., and Tikoo, S. 2004).

Impact of Communication on Organizational Performance

1. Employee Engagement and Retention: Effective communication that maintains high-quality content standards fosters a positive onboarding experience for new hires, leading to higher levels of job satisfaction and employee engagement. When employees understand their roles, expectations, and the organization's goals clearly from the outset, they are more likely to be committed to their jobs and the organization, reducing turnover rates (Welch, M. 2012).
2. Efficiency and Productivity: Clear, accurate, and comprehensive communication reduces the need for clarification and correction, thereby saving time and resources. This efficiency not only applies to the recruitment process but also enhances overall productivity within the organization as employees have a better understanding of their tasks and expectations (Men, L. R., and Stacks, D. W. 2013).
3. Innovation and Problem-Solving: High-quality content in internal communications encourages the exchange of ideas, enhances knowledge sharing, and fosters a culture of innovation. Employees who receive clear, relevant, and comprehensive information are more likely to contribute to problem-solving and innovation processes within the organization (Gibson, C. B., and Gibbs, J. L. 2006).

In summary, the quality of content of messages plays a crucial role in both the recruitment process and the overall performance of an organization. By prioritizing high-quality communication, organizations can attract better talent, enhance employee engagement and satisfaction, and improve productivity and innovation.

- **Group *and team boundaries:*** To establish criteria for membership in groups and teams' cultural values are needed to be considered. Culture defines boundaries for groups and teams and also defines criteria for inclusion in the group and in the team through group goal clarity, group cohesiveness, and managerial support for teamwork.

Group goal clarity refers to the extent to which the goals of a group or team within an organization are clearly defined, understood, and accepted by all its members. This clarity includes knowing what the goals are, understanding how to achieve them, and recognizing the role of each member in reaching these goals. It is a critical factor in group performance because it aligns the efforts and focus of all members toward achieving common objectives (Locke, E. A., and Latham, G. P. 2002).

Impact of Group Goal Clarity on Recruitment Policy:

- Organizations that emphasize clear group goals are likely to attract and select candidates who are goal-oriented and have a strong alignment with the organization's objectives. This alignment can be evaluated during the recruitment process through questions and assessments designed to determine a candidate's fit with the team and organizational goals.
- Recruitment materials and job postings that articulate clear group goals and the significance of these goals within the broader organizational context can attract candidates who are motivated by purpose and clarity in their work roles.

Impact of Group Goal Clarity on Organizational Performance:

- Group goal clarity is associated with higher levels of group and organizational performance. When teams have a clear understanding of their goals, they can better plan and coordinate their efforts, leading to improved efficiency, effectiveness, and productivity (Weldon, E., and Weingart, L. R. 1993).

Another issue to establish criteria for membership in groups and teams' cultural values is group cohesion which refers to the bonds that hold a group together, including the feelings of mutual attraction and commitment among group members. It encompasses the group's unity, the strength of the relationships within it, and the degree to which members are motivated to remain part of the group and work cooperatively toward common objectives. Group cohesion is a multifaceted concept that can influence group processes and outcomes in complex ways (Carron, A. V., and Brawley, L. R. 2000).

Impact of Group Cohesion on Recruitment Policy:

- A focus on group cohesion in recruitment policies can guide the selection of candidates who not only have the necessary skills and qualifications but also are likely to contribute positively to the team dynamics. This can include evaluating

interpersonal skills, cultural fit, and the ability to collaborate effectively with others.
- Organizations that promote a cohesive work environment may emphasize this aspect in their employer branding and recruitment messaging to attract candidates who value teamwork and a supportive work culture.

Impact of Group Cohesion on Organizational Performance:

- High level of group cohesion is associated with enhanced performance, as cohesive groups tend to have better communication, less conflict, and more effective coordination. Members of cohesive teams are more likely to support each other, share knowledge, and work together toward the achievement of group goals, which in turn can lead to higher organizational productivity and success (Beal, D. J., et al. 2003).

In summary, both group goal clarity and group cohesion play crucial roles in shaping the effectiveness of an organization's recruitment policy and its overall performance. By carefully selecting candidates who align with clear group goals and contribute to a cohesive team environment, organizations can enhance their operational effectiveness, improve employee satisfaction, and achieve their strategic objectives.

- ***Power and status:*** Determining the rules for acquiring, maintaining and, losing power and status in an organisation is another important factor in achieving internal integration. The critical issues for the strategy of organization related to power and status are how authority will be allocated to the newcomers and members of the organization with respect to their status and to what extent there is a need for autonomy?

Autonomy is defined as the extent to which a job provides an employee with the discretion and independence to schedule their work and determine the procedures to be used in carrying it out. It is considered a critical psychological state in Hackman and Oldham's Job Characteristics Model, which posits that high levels of autonomy lead to greater job satisfaction, intrinsic motivation, and overall job performance (Hackman, J. R., and Oldham, G. R. 1976).

Impact of Autonomy on Recruitment Policy:

1. Attractiveness to Prospective Employees: Job advertisements that highlight roles offering autonomy are likely to attract candidates who value independence and self-direction. This can be particularly appealing to highly skilled professionals or those in creative and innovative fields where freedom to explore and develop new ideas is valued (Pink, D. H. 2009).
2. Employer Branding: Organizations known for providing autonomy in the workplace can enhance their employer brand, making them more attractive to potential employees. This reputation can help attract a larger pool of candidates, giving the organization a competitive advantage in talent acquisition (Miner, J. B. 2007).

Impact of Autonomy on Organizational Performance:

1. Employee Motivation and Satisfaction: Autonomy is linked to higher levels of job satisfaction and intrinsic motivation. When employees feel that they have control over their work and its outcomes, they are more likely to take initiative, be creative, and persevere through challenges (Deci, E. L., and Ryan, R. M. 2000).
2. Productivity and Innovation: Autonomy can lead to increased productivity and innovation within organizations. Employees with the freedom to experiment and manage their workflows can develop more efficient processes and innovative solutions to problems. This fosters a culture of continuous improvement and adaptability (Amabile, T. M. 1997).
3. Employee Retention: Higher levels of autonomy contribute to reduced turnover rates. Employees who perceive that they have autonomy and control over their work are less likely to experience job burnout and more likely to remain with an organization long-term (Lee, M. D., and MacDermid, S. M. 2017).

In conclusion, autonomy is a key factor in both attracting talent through recruitment policies and enhancing organizational performance through increased motivation, satisfaction, productivity, and innovation. By understanding and implementing autonomy in job design and management practices, organizations can create a work environment that leverages the full potential of their employees.

- ***Rewards and punishments***: To develop systems for encouraging desirable behavior and discouraging undesirable behaviors. To create an attractive image of the organization for the newcomers and to empower members of the organization intrinsic and extrinsic reward satisfaction and performance reward expectancy of employees, reward expectancy of teamwork must be considered as elements of effective reward system instead of punishments in the organization. On the other hand, punishments may carry the risk of discouraging also desirable behaviors of employees and losing attractiveness of the organization in the eye of newcomers and may have a negative impact on internal integration and also performance of the organization.

Rewards satisfaction and performance reward expectancy are fundamental concepts in organizational behavior and human resource management, affecting both the recruitment policies of organizations and their overall performance.

Rewards satisfaction refers to an individual's contentment with the rewards they receive for their job performance, including both tangible rewards (such as pay, bonuses, and benefits) and intangible rewards (such as recognition, professional development opportunities, and work–life balance).

Impact of Reward Satisfaction on Recruitment Policy and Organizational Performance:

- Organizations that offer competitive and satisfying reward packages are more likely to attract high-quality candidates. The perception of rewards offered by an organization can significantly influence a potential employee's decision to apply for a position or accept a job offer (Armstrong, M., and Taylor, S. 2020).

- Rewards satisfaction can lead to higher levels of employee engagement, reduced turnover, and improved job performance. Employees who are satisfied with their rewards are more likely to be committed to their employer and motivated to contribute to the organization's success (Milkovich, G. T., et al. 2019).

The other fundamental concept in organizational behavior and human resource management is performance reward expectancy. This concept is derived from expectancy theory, which suggests that individuals are motivated to perform at certain levels based on the expectation that their efforts will lead to desired outcomes, including receiving certain rewards. This expectancy is a function of the perceived relationship between effort, performance, and rewards.

Impact of Performance Reward Expectancy on Recruitment Policy and Organizational Performance:

- An organization's ability to clearly communicate the link between performance and rewards during the recruitment process can attract candidates who are motivated by such incentives and who have confidence in their ability to meet performance standards (Vroom, V. H. 1964).

Both rewards satisfaction and performance reward expectancy play critical roles in shaping the effectiveness of recruitment policies and the overall performance of an organization. By understanding and strategically managing these aspects, organizations can create a work environment that not only attracts but also retains top talent, thereby enhancing their competitive advantage and operational success.

3.2 Research Questions

We would like to determine the importance of issues dealing with the function of organizational culture from perspectives of managers and employees. Those issues will be the key indicators for the sustainability of corporate management. From those perspectives, the following questions are to be answered:

Which issue or a group of issues has a prime importance for the integration of new employees into the organization? What are the possible impacts of issues on the recruitment strategies and performance of the organization?

The results will reflect the key indicators (elements) for the success and maintenance of internal integration and establishment of the criteria for selection, promotion, and rewards. As it is known, internal integration of the organization can be succeeded and maintained through the effective recruitment process by establishing valid criteria to be used to determine who is assigned to specific jobs or positions, who get raises and promotions, and why, who is removed from the organization. Briefly, effective strategy may be based on the recruitment of employees who fit the culture and removal of employees who deviate from the culture. The objective of our analysis was to extract principal components of issues dealt with internal

Value Label	Value	Frequency	Percent	Valid Percent
GENERAL MANAGER	10.0	1	.5	.5
ASSISTANT GENERAL MANAGER	11.0	6	2.8	2.8
GROUP DIRECTOR	12.0	5	2.3	2.3
DIRECTOR	13.0	13	6.1	6.1
MANAGER	14.0	37	17.4	17.4
DEPARTMAN MANAGER	15.0	94	44.1	44.1
SUPERVISOR	16.0	57	26.8	26.8
	Total	213	100.0	100.0

Valid cases 213 Missing cases 0

Fig. 3.1 Profile of managers in the sample

integration of the organization which affected the strategies of recruitment and performance for enabling sustainable corporate management.

3.3 Method

3.3.1 Sample

Research has been carried out at a company level with separate samples of 213 managers and 841 employees of a leading company in electronics industry. Profiles of managers and employees are shown at Figs. 3.1 and 3.2.

3.3.2 Measures

Following instruments were used to assess the responses of managers and employees on six-point scale.

The reward satisfaction was assessed with two subscales in order to determine satisfaction of the members of organization with certain intrinsic and extrinsic rewards (Seashore et al. 1982).

Performance reward expectancy: Performance reward expectancy is the perceived probability that good performance will lead to rewards (Sims et al. 1976).

Autonomy: The extent to which members of organization have a major say in scheduling their work, selecting the equipment they will use, and deciding on the procedure to be followed (Porter 1961, 1962).

Quality of content of messages: Sufficiency, accuracy, clearness, timeliness, and importance of the messages received by both managers and employees (Porter and

Value Label	Value	Frequency	Percent	Valid Percent
ENGINEER	1.0	310	36.9	36.9
SPECIALIST	2.0	126	15.0	15.0
TECHNICIAN	3.0	111	13.2	13.2
TESTER	4.0	85	10.1	10.1
OPERATOR	5.0	209	24.9	24.9
	Total	841	100.0	100.0

Valid cases 841 Missing cases 0

Fig. 3.2 Profile of employees in the sample

Roberts 1977). Other issues are related to establish a common language in the organization such as frequency and timing of communication and time allocated to communication are also measured.

Group goal clarity and group cohesion: Scales are based on the Michigan Organizational Assessment Questionnaire which contains a number of scales for measuring work attitudes and perceptions. Group goal clarity and group cohesion are subscales of the Module for Work Group Functioning (Cammann et al. 1979). Reward expectancy and managerial support for teamwork are also considered in the analysis.

Cronbach's alpha was the measure of reliability of each instrument used in the research.

The research sought to extract principal components of issues dealt with function of organizational culture affecting the strategies of recruitment. The method applied to the issues in our empirical study is related to exploratory, not confirmatory analysis that is suitable to the purpose of our research. As it is known that exploratory use of a factor-analytic method is the exploration and detection of patterning of variables with a view to the discovery of new concepts and a possible reduction of data.

The measure of sampling adequacy is one of two evaluations that should be performed prior to any attempt to interpret results of principal components analysis. This coefficient quantifies the extent which a composite of variables, and the variables within the composite, conform to the desired expectation of the partial correlations tending toward zero. According to Kaiser (1974), measure of sampling adequacy should be larger than 0.50 in order to assume that Guttman's (1954) assumptions have been minimally met. If the measure exceeds 0.50, research data.

represent a homogeneous collection of variables and are suitable for factor analysis. As the measure approaches one, it is assumed to be that the data are conforming perfectly to the assumptions of zero partial correlations. In our research, sampling adequacy measures for both managers and employees were found to be above 0.50. The results suggested that these data did indeed represent a homogeneous collection of variables and were suitable for factor analysis.

The second evaluation is associated with statistical significance of the correlations. Bartlett's test was applied to the correlation matrix prior to proceeding to factor extraction (Weiss 1970). The result permitted us to reject the null hypotheses for both managers and employees. We concluded that correlation matrices were worth analyzing and represented the factor analysis of meaningful data.

The method applied to the issues in our empirical model is related to exploratory analysis that is suitable to the purpose of our research. As it is known that exploratory use of a factor-analytic method is the exploration and detection of patterning of variables with a view to the discovery of new concepts (Hakstian and Muller 1973).

3.4 Results

The results of reliabilities of instruments used in the analysis are shown in Fig. 3.3.

The correlation matrix was subject to principal components analysis which yielded 3 components for managers and 3 components for employees with eigenvalues greater than one. All components accounted for 60.3% and 63.9% of the total variances in the sample groups of managers and employees, respectively.

Figures 3.4 and 3.5 show the overall result of principal component analysis for managers and employees. Below the label of each factor, composition of variables of each component was stated in order, with respect to magnitudes of their factor loadings.

3.5 Discussion

Communication was perceived by both managers and employees as the most important component dealt with issues of common language and conceptual categories related to internal integration function of organizational culture. Importance of

		Cronbach alpha	
INSTRUMENT	Number of items	managers	employees
intrinsic reward satisfaction	3	.7441	.7290
extrinsic reward satisfaction	3	.6050	.5814
performance-reward expectancy	16	.9236	.9532
degree of autonomy	10	.9083	.8757
quality of content of messages	5	.9022	.9182
group cohesiveness	3	.8314	.8130
group goal clarity	2	.8056	.8262

Fig. 3.3 The results of reliabilities of instruments

Kaiser-Meyer-Olkin Measure of Sampling Adequacy = .75317
Bartlett Test of Sphericity = 642.74916 Significance = .00000

Factor Labels	Factor	Eigenvalue	Pct of Var	Cum Pct
COMMUNICATION	1	3.88405	32.4	32.4
timing of communication				
frequency of communication				
time allocated to communication				
quality of content of messages				
TEAMWORK	2	1.87699	15.6	48.0
group cohesiveness				
managerial support for teamwork				
group goal clarity				
reward expectancy for teamwork				
PERFORMANCE-REWARD Expectancy	3	1.47087	12.3	60.3
performance-reward expectancy				
extrinsic reward satisfaction				
intrinsic reward satisfaction				
degree of autonomy				

Fig. 3.4 Results of principal components analysis for managers

Kaiser-Meyer-Olkin Measure of Sampling Adequacy = .87779
Bartlett Test of Sphericity = 3713.9510. Significance = .00000

Factor Labels	Factor	Eigenvalue	Pct of Var	Cum Pct
COMMUNICATION	1	4.93499	41.1	41.1
time allocated to communication				
frequency of communication				
timing of communication				
quality of content of messages				
PERFORMANCE-REWARD EXPECTANCY	2	1.50160	12.5	53.6
performance-reward expectancy				
extrinsic reward satisfaction				
intrinsic reward satisfaction				
degree of autonomy				
TEAMWORK	3	1.23159	10.3	63.9
group cohesiveness				
group goal clarity				
reward expectancy for teamwork				
managerial support for teamwork				

Fig. 3.5 Results of principal components analysis for employees

communication cannot be only based on the perceptions of members of the organization, it also serves several important functions for an organization. The most important function of communication is to help members of organization to share values which represent collective beliefs, assumptions, and feelings. That is why communication should be considered as a powerful tool for spread of cultural values to all levels of organization.

The primary function of communication is the transmission of core values to new organization members. Newcomers are exposed to these values through the role models they interact with, the training they receive, and the behavior they observe being rewarded or punished. Newcomers are vigilant observers, seeking out clues to the organization's culture and consistency in the cultural messages they receive. If they are expected to adopt these values, it is essential that the message reflect the underlying values of the organization. Communication must be considered as an important vehicle for conveying explicit cultural values to newcomers and employees with the goal to ensure consistent performance and work standards. Research results showed that the most important issue in communication was "time allocated to communication" for employees and "timing of communication" for managers. Transmission of core values through communication enables sustainable corporate management.

Teamwork was another important component which was perceived by both managers and employees. Teamwork has an important function for the internal integration of the organization and has a positive impact on the performance of the organization. Teamwork is considered as a powerful tool for socialization of the newcomers. Teamwork will facilitate their socialization process by mutual relations with other members of the team. Hence, the newcomers can make adjustments in their job and organization to accommodate their knowledge and personalities and can reduce role ambiguity and role conflict which arise from the role demands involving the expectations placed on them. Teamwork is an essential for sustainability of corporate management by reducing role ambiguity and role conflict in the organization. Group cohesiveness was found to be the most important issue in teamwork from the perspectives of both managers and employees. Cohesive groups with norms supportive of management can have several functional consequences for an organization and management. If tasks are interdependent, the cooperative behavior of cohesive group members helps task accomplishment. A cohesive group can produce innovative work behavior that has obvious uses for the organization. And also control over individual behavior in cohesive groups is more immediate than controls used by managers.

Members of the organization perceived the component which was labeled as "performance-reward expectancy" also as an influential factor for organizational culture. An effective rewarding system will enable organization to fulfill its core mission. Newcomers and employees learn about their organization's culture through its reward system. The rewards and punishments attached to various behaviors convey to employees the priorities and values of managers and the organization. A member of an organization must know exactly what behavior is expected in his job. It is possible for him to reach in his work objectives if his role can be clearly defined and if a great deal of feedback and autonomy can be provided. It must be considered that organizational values as important as they are do not have much impact on work performance unless they are reinforced through rewards. That is why organizations must continuously examine their reward systems to find ways to make it more effective in improving employee performance in order to achieve sustainable corporate management.

References

Amabile TM (1997) Motivating creativity in organizations: on doing what you love and loving what you do. Calif Manage Rev 40(1):39–58
Armstrong M, Taylor S (2020) Armstrong's handbook of human resource management practice. Kogan Page
Backhaus K, Tikoo S (2004) Conceptualizing and researching employer branding. Career Dev Int 9(5):501–517
Beal DJ, Cohen RR, Burke MJ, McLendon CL (2003) Cohesion and performance in groups: a meta-analytic clarification of construct relations. J Appl Psychol 88(6):989–1004
Breaugh JA (2008) Employee recruitment. Annu Rev Psychol 59:347–374
Cammann C, Fiehman M, Jenkins D, Kelsh J (1979) The Michigan organizational assessment questionnaire, unpublished manuscript, University of Michigan, Ann Arbor, Michigan
Carron AV, Brawley LR (2000) Cohesion: conceptual and measurement issues. Small Group Res 31(1):89–106
Deci EL, Ryan RM (2000) The "what" and "why" of goal pursuits: human needs and the self-determination of behavior. Psychol Inq 11(4):227–268
Gibson CB, Gibbs JL (2006) Unpacking the concept of virtuality: the effects of geographic dispersion, electronic dependence, dynamic structure, and national diversity on team innovation. Adm Sci Q 51(3):451–495
Guttman L (1954) Some necessary conditions for factor analysis. Psychometrika 19:149
Hackman JR, Oldham GR (1976) Motivation through the design of work: test of a theory. Organ Behav Hum Perform 16(2):250–279
Hakstian R, Muller V (1973) Some notes on the number of factor problem. Multivar Behav Res 8
Kaiser HF (1974) An index of factorial simplicity. Psychometrika 39:31–36
Keyton J (2011) Communication and organizational culture: a key to understanding work experiences. SAGE Publications
Lee MD, MacDermid SM (2017) The impact of work-life balance on employee retention. J Fam Econ Iss 38(4):461–477
Locke EA, Latham GP (2002) Building a practically useful theory of goal setting and task motivation: a 35-year odyssey. Am Psychol 57(9):705–717
Milkovich GT, Newman JM, Gerhart B (2019) Compensation. McGraw-Hill Education
Miner JB (2007) Organizational behavior 4: from theory to practice. M.E. Sharpe
Men LR, Stacks DW (2013) The impact of leadership style and employee empowerment on perceived organizational reputation. J Commun Manag 17(2):171–192
Pink DH (2009) Drive: the surprising truth about what motivates us. Riverhead Books
Porter LW (1961) A study of perceived need satisfactions in bottom and middle management jobs. J Appl Psychol 15:1–10
Porter LW (1962) Job attitudes in management: perceived deficiencies in need fulfilment as a function of job level. J Appl Psychol 46:375–384
Porter LW, Roberts KH (1977) Communication in organizations. Penguin Books Ltd, Middlesex
Schein EH (1988) Organizational culture and leadership. Jossey-Bass, San Fransisco
Seashore SE, Lawler EE, Mirvis P, Cammann C (eds) (1982) Observing and measuring organizational change: a guide to field practice. Wiley, New York
Sims HP, Szilagyi AD, Keller RT (1976) The measurement of job characteristics. Acad Manage J 19:195–212
Weiss DJ (1970) Factor analysis and counselling research. J Couns Psychol
Welch M (2012) The significance of the communication of corporate social responsibility for employee engagement: the example of ING. Public Relat Rev 38(2):277–284
Weldon E, Weingart LR (1993) Group goals and group performance. Br J Soc Psychol 32(4):307–334
Vroom VH (1964) Work and motivation. Wiley

Rauf Nişel, born in 1953, graduated from Boğaziçi University, Faculty of Administrative Sciences in 1977. He received his Master's degree in Management and Organization from Istanbul University, Faculty of Business Administration in 1978. In 1979, he joined the Department of Statistics, Democracy and Economics Analysis at Istanbul University, Faculty of Business Administration, as a reserach assistant. In 1986, he received his doctorate from Istanbul University, Faculty of Business Administration, Department of Quantitative Methods, with his thesis titled Productivity, Cost and Structure, Statistical Analysis of Productivity-Cost Relationships; In 1991, he received the title of Associate Professor in the field of quantitative Methods at Marmara University, and in 2000, he received the title of Professor at the same university and in the same field. Prof. Dr. Rauf Nişel's academic activities cover a wide range. He has given many courses at undergraduate and graduate levels in different fields, such as Business, Engineering, and Medicine Faculties, and at other universities (Istanbul, Marmara, Yeditepe, and Piri Reis Universities). He has supervised numerous research projects and academic theses in these areas. He has many books and national and international articles on quantitative methods. He gave seminars and conducted scientific research at the Universities of Glasgow, Strathclyde, and the Berlin School of Economics and Law.

Chapter 4
Optimizing EV Charging Locations for Full Coverage: D400 and D550 Highways Case in Türkiye

Gamze Neşvat and Eyüp Çetin

Abstract The increasing concern about the environmental damage and sustainability of fossil fuels has led to a surge in electric vehicle adoption. However, the "range anxiety" barrier, which concerns that electric vehicles may not have enough range to reach their destinations, may deter potential buyers. This study aims to create a mathematical model that will help place the fewest possible charging stations needed to allow anyone to travel anywhere within the network, charging their vehicles as needed. The developed model will be applied to the D400 and D550 highways in Turkey, aiming to identify the best regions for investment in energy infrastructure and support Turkey's transition toward an electric-focused future. Addressing this concern is crucial for promoting wider acceptance of electric vehicles.

Keywords Electric vehicle charging stations · Network flow algorithms · Mixed-integer programming (MIP) · Range anxiety · Flow-refueling location problem · Spatial optimization · Accessibility to electric vehicles

---

G. Neşvat
Graduate School of Business, School of Business, Istanbul University, Istanbul, Turkiye

E. Çetin (✉)
Department of Quantitative Methods, School of Business, Istanbul University, Istanbul, Turkey

Laboratory for Industrial and Applied Mathematics, York University, Toronto, ON, Canada

New York Business Global, College Park, MD, USA
e-mail: eyup.cetin@istanbul.edu.tr

K. T. Çalıyurt (ed.), *New Approaches to CSR, Sustainability and Accountability, Volume VI*, Accounting, Finance, Sustainability, Governance & Fraud: Theory and Application, https://doi.org/10.1007/978-981-96-5937-1_4

4.1 Introduction

Transportation serves as the backbone of modern society, enabling individuals to navigate their daily lives with ease and efficiency. Whether it is commuting to work, educational institutions, or various destinations, the reliance on vehicles is undeniably significant. In metropolitan areas like Istanbul, the dependence on transportation is particularly highlighted, with the average daily commute reaching up to 20 km (Masoumi 2020). However, despite its importance, the transportation sector is a significant contributor to environmental issues, accounting for around 20% of global CO2 emissions, with road travel alone responsible for three-quarters of these emissions, with passenger vehicles contributing 45.1% and cargo trucks adding 29.4%, respectively (Climate Watch 2018).

In response to the escalating environmental and sustainability concerns associated with the use of fossil fuels, there has been a significant expansion in the adoption of electric vehicles (EVs) worldwide. The shift toward sustainable transportation is caused not only by increased environmental awareness among individuals but also by systemic changes initiated by governments worldwide, aiming to reduce carbon emissions and decrease dependence on fossil fuels (Burch and Gilchrist 2020). This concerted push toward sustainability has been echoed within the automobile industry, resulting in a remarkable increase in electric vehicle sales (Global EV Outlook 2023—Analysis - IEA 2023).

In November 2023, electric vehicles (EVs) accounted for 12.1% of all brand-new automobile sales in Turkey, a significant increase from the 2.7% recorded in the previous year. This growth in EV usage indicates a shift in consumer preferences toward greener, more sustainable transportation. Moreover, Tesla, a leading EV manufacturer, initiated direct sales in Turkey in 2023, enhancing accessibility for Turkish consumers. Turkey's long-held goal of creating a national auto brand was realized with the launch of the TOGG brand. This initiative, solely focused on the production of an electric SUV, signifies a brave step into the electric car industry. The combination of Tesla's entry and TOGG's debut plays a crucial role in facilitating Turkey's shift toward a transportation paradigm that is more sustainable and environmentally friendly.

Although electric vehicles (EVs) offer numerous benefits, a significant barrier that often discourages potential buyers is known as "range anxiety" (Rauh et al. 2014). This term describes the worry that an EV may not have sufficient range to reach its desired destination, raising the fear of being stranded before arriving. EVs are best for short urban commutes, but Turkey's vast geographical landscape makes it difficult for them to handle lengthy intercity trips without numerous charging stops. Additionally, when operated at high speeds on highways, EVs tend to exhibit a diminished actual range compared to the manufacturer's stated range, further worsening range anxiety (Jonathan and Thomas 2016). Effectively addressing widespread concerns is crucial for promoting the acceptance and adoption of electric vehicles, as it directly influences prospective buyers' confidence in EVs as reliable travel alternatives.

Turkey's investment in the national electric vehicle brand TOGG and support for the Glasgow Declaration serve as indicators of its dedication to a sustainable future. This declaration, signed at the UN Climate Change Conference in Glasgow in 2022, sets an ambitious goal to cease the sale of new diesel and petrol vehicles by no later than 2040 (The UN Climate Change Conference in Glasgow 2022). This landmark agreement not only highlights the global shift toward more sustainable transportation options but also underscores the urgency of transitioning away from fossil fuels. Given that electric vehicles are at the forefront of this transition, the need for a robust and efficient charging infrastructure has never been more critical. Ensuring the availability of such infrastructure is paramount to supporting the growing adoption of electric vehicles, thereby reinforcing Turkey's commitment to a greener, more sustainable future and addressing one of the main obstacles in the global fight against climate change.

Given the emerging landscape shaped by these advancements, we are motivated to design and refine a mathematical model that strategically pinpoints the optimal locations for installing electric vehicle (EV) charging stations. The core aim of this model is to facilitate seamless travel for EVs across varying range capacities, ensuring drivers can access charging services as needed without encountering range anxiety or significant detours. This model incorporates critical variables, including population density and average vehicular speed, to optimize charging station placements effectively.

The insights generated by our model highlight regions that may experience increased energy demands, allowing for targeted investments in energy infrastructure enhancements. By addressing the infrastructural needs for EV adoption, we aim to advance Turkey's transition to sustainable transportation, aligning with global environmental goals and providing a framework for other nations to follow.

4.2 Related Literature

The pursuit of finding optimal sites for electric vehicle (EV) charging stations has sparked widespread interest. While some studies focus on optimizing slow charging stations—which are intended for overnight use—others focus on fast-charging stations, which are made for quick top-ups that are necessary for long-distance travel.

Frade et al. (2011), using population data for each district in Lisbon, predicted night and day charging requirements and then applied the highest coverage model to determine where charging stations could best meet the expected demand. Chen et al. (2013) chose parking demand as a suitable indicator for charging station demand and implemented a mixed-integer programming problem (MIP) in Seattle, Washington, aimed at minimizing access costs for vehicle owners. Xiang et al. (2017) used GDP, population, subsidies, competitive vehicle prices, and charging times in their models. Bayram et al. (2015) enabled charging stations to serve various types of customers, taking into account limited-service capacity constraints and possible congestion, and aimed at minimizing charging times and energy needs from the local power grid. These models, using secondary data sources, have tried to solve the problem of the location of non-fast-charging stations (e.g., vehicles that are left off at night) and have been useful for periods when electric vehicles are not yet widespread.

The models to be developed for long distances, unlike the above models, are set up to take into account the range problem. Kuby and Lim (2005) developed the first flow-refueling location problem (FRLM) to optimally position fuel storage facilities for vehicles with limited range. The aim of the FRLM is to maximize the flow of traffic that can be completed through a fixed number of fuel charging stations (or electric vehicle charging, which can be applied with the same logic). Many studies later developed the FRLM model for the various goals. Capar et al. (2013) developed the FRLM model and managed to reduce the resulting time of the model by 70%. Wang and Lin (2009) suggested the improved FRLM model, which attempts to reduce the costs of building or maintaining fueling stations while supplying fuel to all of the vehicles. Lee et al. (2014) developed a model on the FRLM in 2014 to locate the fast-charging stations of electric vehicles, taking into account the charging status of batteries and users' charging and travel habits. This model aims to place charging stations in a way that does not adversely affect traffic flow, taking into account traffic congestion in urban areas.

MirHassani and Ebrazi (2013) made it feasible to use FRLM in more realistic networks by proposing a preprocessing network expansion technique that needs to be carried out prior to main modeling. This improvement makes it easier to work on large data sets. Chung and Kwon (2015) formulated an optimization model based on a flow-refueling location model (FRLM) to maximize network coverage along South Korean highways.

This study, based on the FRLM model developed by (MirHassani and Ebrazi 2013), identifies an opportunity to use data published by the Directorate-General of Highways of the Ministry of Transport and Infrastructure of the Republic of Turkey. The model incorporates average speed data to calibrate the range of drivers on highways where they operate at elevated velocities. Population density is utilized to enhance the effectiveness of the model.

4.3 Materials and Methods

The methodology of this study is based on the assumption that charging stations need to be placed in such a way that any electric vehicle can reach the next charging station before running out of charge. To determine the optimal placement of charging stations, the study takes into account factors such as the average range of electric vehicles, the distance between points of interest, and the average speed. By considering these variables, the model aims to ensure that drivers have convenient access to charging infrastructure and can confidently travel long distances without worrying about running out of charge.

Paths and connections are mathematically modeled using graph theory. In this context, the nodes indicate the defined points in the path, whereas the connections (arc) indicate the road between these points (Seker 2015).

The model will be constructed using the "flow-refueling location model" (FRLM) developed by Kuby and Lim (2005) and the models built on the FRLM by MirHassani and Ebrazi (2013). The model consists essentially of two stages. In the first phase, all possible combinations that can be used for charging will be created, and then charging stations will be placed using mixed-integer programming.

4.3.1 Data Collection

The initial phase of the process involves gathering the requisite data from suitable sources. The data utilized mostly consisted of industrial data provided by government entities and firms.

4.3.1.1 Ranges of Various Electric Vehicle Models

Range is a critical value that indicates how far an electric vehicle can travel before needing to recharge. The range of an electric vehicle is based on several different factors, such as battery size, driving conditions, and the driver's habits. To find the ranges of electric vehicles, the WLTP range data published by their manufacturers will be used. The Worldwide Harmonized Light Vehicle Test Procedure (WLTP) is a global standard for determining the levels of pollutants, CO2 emissions, and fuel consumption of conventional and hybrid cars and also determining the range of electric vehicles (Worldwide Harmonised Light Vehicle Test Procedure 2021). The WLTP range values of the 10 best-selling electric vehicles in Turkey in 2023 are presented in Table 4.1. The range value to be used as an input to the model has been

Table 4.1 Electric vehicle sales rates and ranges in January–October 2023

Electric vehicle model	Sell-through rate	WLTP range value (km)
Tesla Model Y	32,34%	455
TOGG T10X V2	27,72%	314
MG4	6,55%	435
Renault Megane E-Tech	6,33%	431
Skywell ET5	6,17%	460
Renault Zoe	5,26%	395
Opel Mokka	4,04%	327
Dacia Spring	3,24%	310
Volvo XC40	3.08%	498
MG ZS EV	3.00%	440

Source: (NTV 2023)

determined to be 310 km, which is the range of the vehicle with the lowest range (Dacia Spring) in the best-selling electric vehicles table. By using the range value of the vehicle with the lowest range, the model can accommodate a wide range of electric vehicles in Turkey. This approach allows for a comprehensive analysis of the WLTP range values and ensures that the model is applicable to various electric vehicle models on the market.

4.3.1.2 Geographical Road Data

Geospatial road data is the basis for building and operating our model, which is designed to identify potential charging station locations. The General Directorate of Highways publishes a report covering road lengths, traffic densities, and average speed data on an annual basis. The data required from this report will be processed using data mining techniques.

Our initial step involved a detailed examination of road segment data, filtered from comprehensive reports by the General Directorate of Highways. This report systematically organizes road segments, each tagged with unique identifiers, for example, "400-01" for the initial 49-km stretch of the D400 in Muğla. Since the regional directorates, rather than the road numbers, were used to group the data in the General Directorate of Highways report, we created a Python script to extract all tabular data from PDF files into a CSV format, making it easier to intercept.

To simplify the model building process, the focus was on two main roads: the D550, the longest state road in Turkey, and the D400, the longest state road in Turkey in an east-west direction. One of the reasons for choosing the D400 and D550 roads is that these roads pass through important summer tourism spots in Turkey, especially the west and south coasts. These roads are heavily used during the peak tourist season for intercity travel, which makes them an ideal option for installing charging stations. The model makes it easy to determine the exact locations for charging stations on these roads, thus ensuring a smooth, uninterrupted journey for electric vehicles. In addition, the scalability of the model is not limited in these specific ways. Using the same tools and data from the other roads and connecting routes mentioned in the report, the model can be easily implemented across the country. This wider application potential ensures that the model has a wide-ranging impact beyond the pathways initially chosen.

The General Directorate of Highways extracted the data pertaining to the D400 and D550 highways for the report, as indicated in Tables 4.2 and 4.3. As shown in Fig. 4.1, these two roads consist of a total of 160 road sections and have a total length of 2801 km (Traffic and Transportation Information for 2022: Annual Average Daily Traffic Values and Transportation Information of Highways and State Roads by Traffic Zones 2023). The intersections of the road sections are chosen as potential charging station installation points. In addition, since the longest stretch of road is 29 km, all electric vehicles can travel between any two points without the need for recharging.

Table 4.2 D550 road segments

No	Segment	Length (km)	Avg. Vehicle	Avg. Speed	Start	End
550-01	1	18	6.802	107	Havsa	Kırcasalih
550-01	2	19	6.632	92	Kırcasalih	Uzunköprü
550-01	3	19	8.463	97	Uzunköprü	İpsala k.
550-02	1	26	5.299	101	İpsala k.	Malkara k.
550-02	2	2	19.470	76	Malkara k.	Keşan
550-02	3	22	6.836	102	Keşan	Çanakkale s.
550-03	1	18	6.413	97	Çanakkale s.	Şarköy k.
550-03	2	29	7.297	104	Şarköy k.	Gelibolu k.
550-03	3	38	5.177	95	Gelibolu k.	Eceabat k.
550-04	1	20	22.286	85	Eceabat k.	Eceabat
550-04	2	12	12.985	106	Eceabat	Çanakkale
550-04	3	17	10.033	104	Çanakkale	İntepe
550-05	1	20	8.319	104	İntepe	Bozcaada k.
550-05	2	29	6.553	103	Bozcaada k.	Ezine
550-06	1	33	34.455	89	Ezine	Ayvacık
550-06	2	3	34.430	69	Ayvacık	Balıkesir s.
550-07	1	10	22.688	89	Balıkesir s.	Edremit
550-07	2	18	11.672	69	Edremit	Balıkesir k.
550-07	3	6	15.033	87	Balıkesir k.	Burhaniye
550-07	4	10	15.004	99	Burhaniye	Gömeç
550-07	5	11	14.348	91	Gömeç	Kozak k.
550-07	6	4	9.890	93	Kozak k.	Ayvalık
550-08	1	18	12.222	98	Ayvalık	Altınova
550-08	2	17	16.057	100	Altınova	İzmir s.
550-08	3	16	16.414	103	İzmir s.	Dikili
550-09	1	23	16.141	87	Dikili	Çandarlı
550-09	2	8	27.082	83	Çandarlı	Çandarlı k.
550-09	3	7	26.282	87	Çandarlı k.	Aliağa
550-09	4	11	30.141	82	Aliağa	Foça kuzey k.
550-09	5	4	40.912	73	Foça kuzey k.	Foça güney k.
550-10	1	7	38.129	85	Foça güney k.	Menemen
550-10	2	7	27.062	80	Menemen	Menderes
550-10	3	5	35.816	73	Menderes	Ayrancılar
550-10	4	7	15.372	82	Ayrancılar	Torbalı k
550-11	1	11	7.340	96	Torbalı k	Torbalı
550-11	2	12	14.076	91	Torbalı	Tire k.
550-11	3	8	6.223		Tire k.	Belevi
550-11	4	8	7.317	96	Belevi	Selçuk
550-12	1	5	7.818		Selçuk	Kuşadası k.
550-12	2	8	13.552	81	Kuşadası k.	Aydın s.
550-12	3	11	13.048	86	Aydın s.	Ortaklar

(continued)

Table 4.2 (continued)

No	Segment	Length (km)	Avg. Vehicle	Avg. Speed	Start	End
550-12	4	8	26.986	78	Ortaklar	Germencik
550-13	1	6	19.317	93	Germencik	İncirliova
550-13	2	21	19.329	102	İncirliova	Aydın
550-13	3	7	17.717	67	Aydın	Koçarlı k.
550-13	4	22	14.576	100	Koçarlı k.	Karpuzlu k.
550-14	1	12	10.941	109	Karpuzlu k.	Çine
550-14	2	3	18.149		Çine	Muğla s.
550-14	3	19	22.941	99	Muğla s.	Kavaklıdere k.
550-14	4	9	18.859	96	Kavaklıdere k.	Yatağan
550-14	5	10	22.003		Yatağan	Yeşilyurt k.
550-14	6	13	16.257	94	Yeşilyurt k.	Muğla
550-15	1	3	4.967	61	Muğla	Ula k.

Source: (Traffic and Transportation Information for 2022: Annual Average Daily Traffic Values and Transportation Information of Highways and State Roads by Traffic Zones 2023)

4.4 Data Preparation

In the data preparation phase of this study, we initially identified 161 potential locations for the placement of electric vehicle (EV) charging stations along the D400 and D550 highways. However, to address performance constraints and ensure the scalability of our model, we refined this initial list to a strategic selection of 57 sites. This selection was based on a careful analysis of population density and traffic volume, aiming to position charging stations in areas that would best serve the expected demand.

For the purpose of optimizing the network, we explored all possible direct routes emerging from these 57 sites, resulting in a comprehensive set of 1596 unique pairs of start and target points. To refine our analysis further, we assessed these routes based on the criterion that any start–target pair separated by a distance less than half of our 310-km range benchmark (i.e., 155 km) would likely not require a mid-journey recharge. This assumption allowed us to exclude such pairs from our optimization model, narrowing our focus to 1195 unique start–target pairs deemed essential for coverage analysis.

This refinement process was critical in concentrating our efforts on identifying where charging stations are most needed, ensuring that EV users embarking on longer journeys across the D400 and D550 highways have access to the necessary charging infrastructure. The strategic focus on these 1195 pairs helps us direct our resources and planning toward areas that, according to our analysis, will maximize the utility and accessibility of the charging network for EV users traveling these key routes.

Table 4.3 D400 road segments

No	Segment	Length (km)	Avg. Vehicle	Avg. Speed	Start	End
400-03	2	18	12.974	84	Ula k.	Köyceğiz
400-03	3	18	17.240	96	Köyceğiz	Ortaca
400-03	4	4	21.974		Ortaca	13. Bölge sınırı
400-03	5	1	22.403		13. Bölge sınırı	Dalaman k.
400-04	1	11	13.127	97	Dalaman k.	Göcek
400-04	2	30	16.654	93	Göcek	Fethiye
400-05	1	20	18.710	90	Fethiye	Seydikemer
400-05	2	22	8.072	87	Seydikemer	Eşen
400-05	3	13	5.265		Eşen	Antalya s.
400-06	1	23	7.711	76	Antalya s.	Kalkan
400-06	2	24	5.273	75	Kalkan	Kaş
400-06	3	10	5.270	73	Kaş	Elmalı k.
400-07	1	37	2.723	88	Elmalı k.	Demre
400-07	2	27	4.370	67	Demre	Finike
400-07	3	2	14.136		Finike	Elmalı k. 2
400-08	0	16	14.136	86	Elmalı k. 2	Kumluca
400-09	1	36	8.124	100	Kumluca	Tekirova
400-09	2	12	12.839	89	Tekirova	Kemer
400-09	3	14	22.293	89	Kemer	Beldibi
400-09	4	10	22.335	83	Beldibi	KGM 13. Bölge
400-09	5	11	24.731	91	KGM 13. Bölge	Antalya Limanı
400-09	6	6	32.895	83	Antalya Limanı	Antalya k.
400-09	7	8	17.821	96	Antalya k.	Antalya
400-10	1	10	46.159	85	Antalya	Belek
400-10	2	5	48.070	84	Belek	Serik
400-10	3	10	33.111	95	Serik	İbradi k.
400-11	1	26	35.563	100	İbradi k.	Manavgat
400-11	2	12	32.053	73	Manavgat	Akseki k.
400-12	1	17	24.941	98	Akseki k.	Alarahan k.
400-12	2	17	27.310	88	Alarahan k.	Gündoğmuş k.
400-12	3	14	34.285	85	Gündoğmuş k.	Alanya
400-12	4	11	31.075	61	Alanya	Sarıveliler k.
400-12	5	11	16.335	88	Sarıveliler k.	Demirtaş k.
400-13	1	21	13.381	89	Demirtaş k.	Gazipaşa
400-13	2	15	8.121	82	Gazipaşa	Kırahmetler
400-13	3	23	4.279	93	Kırahmetler	Mersin s.
400-14	0	39	4.117	70	Mersin s.	Anamur
400-15	1	13	10.148	80	Anamur	Bozyazı
400-15	2	37	4.842	76	Bozyazı	Aydıncık
400-15	3	2	9.665		Aydıncık	Gülnar k.
400-16	1	24	4.935	90	Gülnar k.	Büyükeceli

(continued)

Table 4.3 (continued)

No	Segment	Length (km)	Avg. Vehicle	Avg. Speed	Start	End
400-16	2	38	10.557	94	Büyükeceli	Taşucu
400-16	3	11	19.549	85	Taşucu	Silifke
400-17	1	10	19.583	90	Silifke	Atayurt
400-17	2	10	15.286	76	Atayurt	C.Cehennem
400-17	3	12	16.187	93	C.Cehennem	Kanlıdivane
400-17	4	13	28.968	83	Kanlıdivane	Erdemli
400-17	5	16	36.949	65	Erdemli	Mersin
400-18	1	3	30.055		Mersin	Serbest Bölge
400-18	2	21	25.589	80	Serbest Bölge	Tarsus
400-18	3	4	26.049		Tarsus	Pozantı k.
400-18	4	7	16.027	86	Pozantı k.	Tarsus doğu
400-18	5	10	15.826	93	Tarsus doğu	Adana s.
400-19	1	7	19.754	89	Adana s.	Karaisalı k.
400-19	2	17			Karaisalı k.	Adana
400-19	3	12			Adana	Göztepe
400-19	4	12	15.182	89	Göztepe	Misis
400-19	5	19	21.498	94	Misis	İmamoğlu k.
400-20	1	6	15.259	93	İmamoğlu k.	Ceyhan k.
400-20	2	22	14.287	95	Ceyhan k.	Osmaniye s.
400-21	1	3	6.735		Osmaniye s.	Hatay k1.
400-21	2	3	6.288	88	Hatay k1.	Hatay k2.
400-21	3	16	17.886		Hatay k2.	Osmaniye
400-21	4	27	9.098	89	Osmaniye	Arıklıkaş
400-21	5	6	7.185	86	Arıklıkaş	Bahçe
400-21	6	10	5.642	72	Bahçe	Gaziantep s.
400-22	1	9	3.800	70	Gaziantep s.	Nurdağı
400-22	2	17	6.385	83	Nurdağı	Sakçagöz
400-22	3	37	8.541	77	Sakçagöz	K.maraş k.
400-22	4	10	48.389		K.maraş k.	Karpuzkaya
400-22	5	6	67.275		Karpuzkaya	Gaziantep
400-23	1	5	23.004		Gaziantep	Gaziantep doğu
400-23	2	34	17.757	93	Gaziantep doğu	Nizip
400-23	3	14	9.871		Nizip	Şanlıurfa s.
400-24	1	2	14.550		Şanlıurfa s.	Birecik
400-24	2	4	13.688	87	Birecik	Halfeti k.
400-24	3	39	6.593	78	Halfeti k.	Suruç k.
400-25	1	24	8.678	82	Suruç k.	Ş.urfa batı
400-25	2	16	7.988	85	Ş.urfa batı	Şanlıurfa
400-26	1	15	16.131	95	Şanlıurfa	Akçalı
400-26	2	5	15.527	95	Akçalı	Otoyol k.
400-26	3	46	14.198	98	Otoyol k.	Siverek k.
400-27	1	22	12.155	106	Siverek k.	Viranşehir

(continued)

Table 4.3 (continued)

No	Segment	Length (km)	Avg. Vehicle	Avg. Speed	Start	End
400-27	2	23	11.598	91	Viranşehir	Mardin s.
400-28	1	21	9.608	92	Mardin s.	Derik s.
400-28	2	31	9.449	89	Derik s.	Kızıltepe
400-29	1	25	11.445	89	Kızıltepe	Mardin k.
400-29	2	18	9.236	89	Mardin k.	Akarsu k.
400-30	1	11	9.815	100	Akarsu k.	Nusaybin
400-30	2	7	9.597		Nusaybin	Nusaybin S.K.
400-30	3	38	7.824	90	Nusaybin S.K.	Şırnak s.
400-31	1	36	6.482	110	Şırnak s.	Dicle k.
400-31	2	17	9.312	85	Dicle k.	İdil k.
400-31	3	3	11.650	71	İdil k.	Cizre
400-32	1	31	5.715	81	Cizre	Siirt k.
400-32	2	8	7.469	64	Siirt k.	Şırnak
400-32	3	42	2.552		Şırnak	Uludere k.
400-33	1	14	1.953		Uludere k.	Ortabağ k.
400-33	2	44	1.207		Ortabağ k.	Hakkari s.
400-34	1	30	554		Hakkari s.	Çukurca k.
400-34	2	22	781		Çukurca k.	Ayvalık k.
400-35	0	31	1.165		Ayvalık k.	Hakkari
400-36	1	7	3.628	68	Hakkari	Çukurca k. 2
400-36	2	36	1.426		Çukurca k. 2	Van k.
400-36	3	32	2.136		Van k.	Akçalı 2
400-36	4	2	3.864	78	Akçalı 2	Yüksekova
400-37	0	41	1.578	98	Yüksekova	Esendere

Source: (Traffic and Transportation Information for 2022: Annual Average Daily Traffic Values and Transportation Information of Highways and State Roads by Traffic Zones 2023)

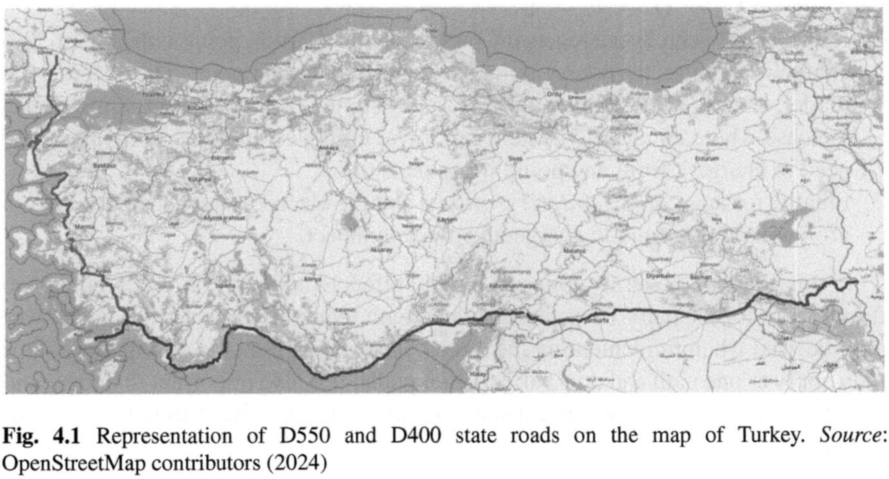

Fig. 4.1 Representation of D550 and D400 state roads on the map of Turkey. *Source*: OpenStreetMap contributors (2024)

4.4.1 Model Preparing

In the first step, the shortest paths between the selected pairs of points are found using the shortest path finding algorithms, and the points through which these paths pass are recorded. The shortest path between any different start-destination pair is called q, and the set of all combinations of these shortest paths are called Q.

To understand how to model works, at first it will be assumed that there is only one path in existence. $G = \left(\overline{N^q, A^q}\right)$ defines the extended network. N^q represents points (nodes) in the existing network, while $\overline{N^q}$ is defining nodes in the extended network. Similarly, A^q identifies existing path segments (arches), while A^q identifies arcs in the extended network (MirHassani and Ebrazi 2013). $aDist_q(i,j)$ returns the adjusted shortest distance between and points i ve j in the path q. The adjusted distance is a calculated by using a formula which is modeled by analyzing 350,000 trips from 500 sedans, totaling over 180,000 h of driving (How Temperature and Speed Impact EV Range | Geotab, 2023).

$$aDist = distance * \left[0.9 + 2.81 * \left(\frac{1}{v}\right) + 6.04 * 10^{-5} * v^2\right]$$

$ord_q(i)$, returns the order number of node i in the path of q. So, for example, if a path from A to E proceeds in the form of A → B → C → D → E, B's order is 2, and C's order is 3. Now let us look at how to create the extended network step by step (MirHassani and Ebrazi 2013):

1. Two new nodes are added: s as a source and t as a sink. The start node connects to newly added source node and the destination node connects to sink node. The example given above turns into the following for the extended network: s → A → B → C → D → E → t.
2. If an electric vehicle can reach any point from the starting point with at least 50% charge remaining, the source point s is connected to that point. This informs the model, theoretically, that the journey can be started from that point. Similarly, in the case where an electric vehicle can travel to its destination point without the need to charge it from any point, it connects to t.
3. Finally, for any i or j; if $ord_q(i) < ord_q(j)$ and the electric vehicle can travel from i to j without the need to charge, the points connect.

In this context, the term "connect" simply refers to adding new road segments to the network for the path. These newly added path segments are not physical but will only be used when implementing the model. Also, changes made during these steps only apply to one path and don't affect other paths. It is also important to understand that the methods of determining distances and sequences give different results for each path.

To determine the optimal path between node s (source) and node t (destination) through the optimization model, point s is assigned a virtual supply value of 1 and point t is assigned a demand value of −1. The values of all the remaining points are assigned as 0 (MirHassani and Ebrazi 2013). The assignment of these virtual supply and demand values effectively transforms the problem into one of balancing the flow within the network, ensuring that what flows out of the source eventually flows into the destination. This modeling approach allows for the application of network flow algorithms that can efficiently compute the optimal path by considering various constraints, such as capacity limits on the edges connecting the nodes. By setting the intermediary nodes' values to 0, the model emphasizes that these nodes neither generate nor consume the flow but merely serve as conduits through which the flow is routed. This simplification focuses the optimization efforts on efficiently directing the flow from the source to the destination, taking into account the network's structure and constraints to find the most effective path.

4.5 The Model

Parameters:

c_i: The cost of building a station. To simplify the model, the cost of each location is considered fixed 1.

Decision Variables:

y_i: If there is a charging station at the point, it takes the value 1, if not, it takes the value 0.

$x^q_{\{ij\}}$: returns the flow value in the road segment (i,j). In other words, it shows which road segment the electric vehicle prefers.

Model:

$$minZ = \sum_{i\in N} c_i y_i \,\#(1)\; \sum_{\{j|(i,j)\in A^q\}} x^q_{ij} - \sum_{\{j|(j,i)\in A^q\}} x^q_{ji} = \{1 i = s, -1 i = t, 0 i \neq s, t,$$
$$\forall q \in Q, \forall i \in \widehat{N}^q \,\#(2)\; \sum_{\{j|(j,i)\in A^q\}} x^q_{ji} \leq y_i \forall i \in N, \forall q \in Q_i,$$
$$\#(3)\; x^q_{ij} \geq 0 \forall q \in Q, \forall (i,j) \in \widehat{A}^q, \#(4)\; y_i \in \{0,1\} \forall i \in N \#(5)$$

The objective function (1) of the model tries to minimize the number of charging stations to be installed. The first constraint function (2) ensures that there is an available flow for each path by using the available path segments. The restriction function (3) only allows the flow to pass through, if there is a charging station at that point. The constraint function (4) states that the flow cannot be negative, and the constraint function (5) states that the decision to establish a station is a binary variable (MirHassani and Ebrazi 2013).

4.6 Application

For the solution of the optimization model, the application named "OR-Tools Constraint Programming Satisfiability Problem Solver" published by Google AI was used. This tool is an effective model solver application that was chosen because of its ability to handle complex calculations and models and easy to scale by using Google Cloud infastrucure.

When the preprocessing of the model is executed with selected roads, points, and a benchmark range value of 310 km, it turns into a mixed-integer problem with 274,234 constraints and 1,418,748 variables. These large numbers emphasize both the complexity and detail of the problem.

The problem was run on the Macbook Pro M1 Pro computer, and the running time lasted approximately 20 min and 55 s. As a result, as can be seen in Table 4.4 and Fig. 4.2, the model shows that if only 14 points are selected as charging stations, an electric vehicle that will use any start-destination point on this road can reach the destination point and return to the back starting point without being on the road, provided that it recharges where necessary. This result also optimizes the use of resources and infrastructure, contributing to the viability and efficiency of electric vehicle use in this region.

(Range = 310 km)

To go deeper into the impact of changing the benchmark range, we reran our model for distances of 100, 150, 200, and 250 km. Considering that the majority of an electric vehicle's (EV) cost is due to its battery (König, A. et al., 2021), it is important to explore the possibility of producing EVs with smaller batteries. We also need to figure out how many more charging stations would be necessary to support these vehicles.

Table 4.4 Location names and coordinates of 14 charging stations to be placed according to the result of the model (Range = 310 km)

Location	Coordinates
Gelibolu kavşağı	40.422604, 26.669222
Ayvacık	39.598931, 26.432040
Foça güney kavşağı	38.685147, 27.008275
Karpuzlu kavşağı	37.662015, 28.020270
Fethiye	36.662310, 29.147475
Demre	36.238815, 29.988762
Serik	36.912408, 31.105906
Mersin—Antalya Sınırı	36.107147, 32.564746
Silifke	36.375215, 33.937607
Ceyhan kavşağı	37.051572, 35.833830
Gaziantep	37.083882, 37.424471
Akçalı	37.699836, 44.006056
Nusaybin	37.090518, 41.212272
Ayvalık köy yolu	37.465923, 43.509758

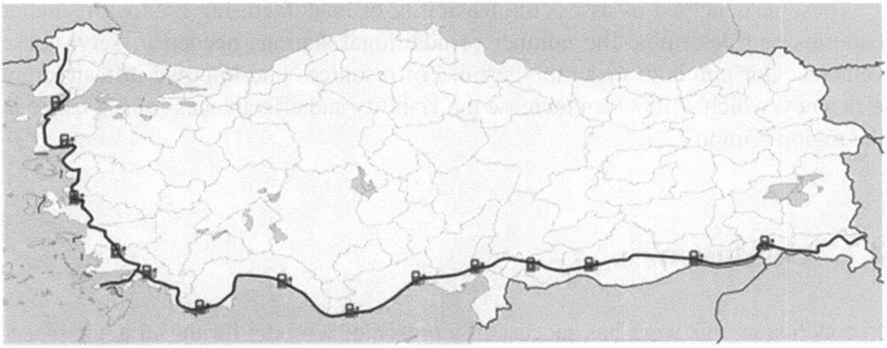

Fig. 4.2 14 charging stations to be placed according to the result of the model

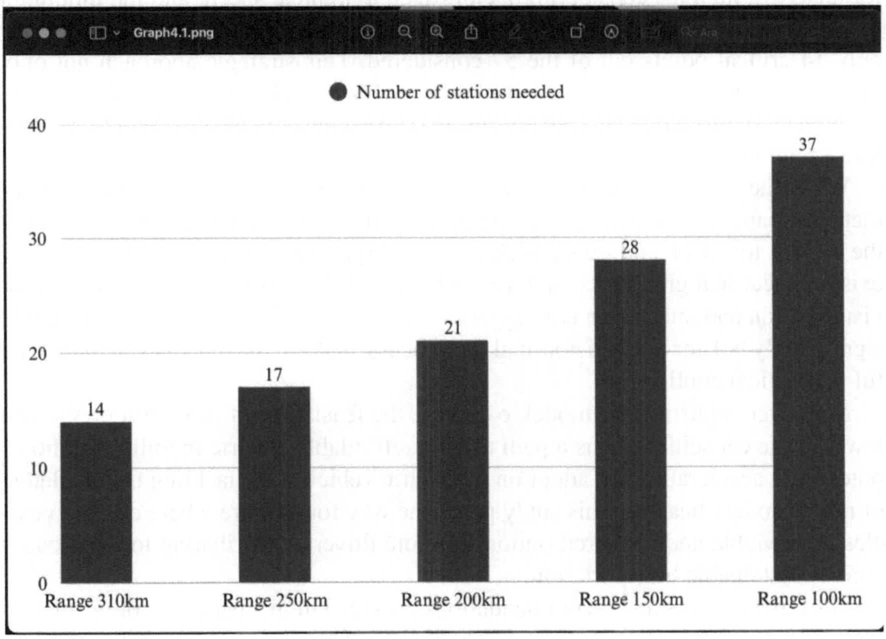

Graph 4.1 Number of stations needed by different ranges. *Source*: OpenStreetMap contributors (2024)

As shown in Graph 4.1, reducing the range to 250 kilometers required three additional stations. Further reducing the range to 200 kilometers still only required three more stations. However, if we were to reduce the range to 100 kilometers, we would need 23 more stations compared to our initial 310-kilometer range.

These tests helped us assess the feasibility of manufacturing EVs with smaller batteries and determine the number of additional stations needed to serve these vehicles. Our aim is to make the best use of resources and improve infrastructure efficiency, which will in turn increase the viability and effectiveness of EV usage in the region (Graph 4.1).

4.7 Conclusion

In conclusion, this work has successfully presented a model for the strategic placement of charging stations along Turkey's D400 and D550 highways, addressing one of the key concerns for electric vehicle (EV) users, which is the availability of charging options during travel. The model has adeptly identified optimal locations for charging stations, considering factors such as vehicle speed, and the minimum range of popular electric vehicles, demonstrating its effectiveness with proposals for only 14 critical points out of the 57 considered. This strategic approach not only promises to enhance the driving experience for EV users by minimizing disruptions due to battery discharge but also ensures that investments in the charging infrastructure are utilized efficiently.

While the current model serves as a robust foundation for infrastructure development, the paper also underscores potential areas for further enhancement. Expanding the model to cover additional highways, incorporating detailed analyses of the existing electrical grid, costs, and user behavior, and integrating advanced tools for visualization and interaction are suggested as next steps. These improvements could significantly bolster the decision-making process, making the model even more useful in practical applications.

Moreover, adjusting the model to explore the feasibility of supporting EVs with lower range capacities opens a path to more affordable electric mobility solutions, potentially accelerating the adoption of electric vehicles. By tackling the challenge of range anxiety head-on, this study paves the way for a future where electric vehicles are a viable and preferred option for more drivers, contributing to the broader goal of sustainable transportation.

Thus, the insights and recommendations provided in this paper not only contribute to the advancement of electric vehicle infrastructure but also underscore the importance of strategic planning in overcoming the barriers to EV adoption. It is a testament to the significance of intelligent infrastructure development in promoting a greener, more sustainable future.

References

Bayram IS, Tajer A, Abdallah M, Qaraqe K (2015) Capacity planning frameworks for electric vehicle charging stations with multiclass customers. IEEE Trans Smart Grid 6(4):1934–1943. https://doi.org/10.1109/tsg.2015.2406532

Burch, Gilchrist (2020) Survey of global activity to phase out internal combustion engine vehicles. The Climate Center. Retrieved February 7, 2024, from https://theclimatecenter.org/wp-content/uploads/2020/03/Survey-on-Global-Activities-to-Phase-Out-ICE-Vehicles-update-3.18.20-1.pdf

Capar I, Kuby M, Leon VJ, Tsai YJ (2013) An arc cover–path-cover formulation and strategic analysis of alternative-fuel station locations. Eur J Oper Res 227(1):142–151. https://doi.org/10.1016/j.ejor.2012.11.033

Chen TD, Kockelman KM, Khan M (2013) Locating electric vehicle charging stations. Trans Res Record J Trans Res Board 2385(1):28–36. https://doi.org/10.3141/2385-04

Chung SH, Kwon C (2015) Multi-period planning for electric car charging station locations: a case of Korean expressways. Eur J Oper Res 242(2):677–687. https://doi.org/10.1016/j.ejor.2014.10.029

ClimateWatch(2018)*DataExplorer|ClimateWatch.*DataExplorer|ClimateWatch.RetrievedFebruary 7, 2024, from https://www.climatewatchdata.org/data-explorer/historical-emissions?historical-emissions-data-sources=cait&historical-emissions-gases=co2&historical-emissions-regions=All%20Selected&historical-emissions-sectors=total-including-lucf%2Ctransportation&page=1&sort_col=country&sort_dir=ASC

Frade I, Ribeiro A, Gonçalves G, Antunes AP (2011) Optimal location of charging stations for electric vehicles in a neighborhood in Lisbon, Portugal. Trans Res Record J Trans Res Board 2252(1):91–98. https://doi.org/10.3141/2252-12

Global EV Outlook 2023—Analysis - IEA (2023) Global EV outlook 2023—analysis—IEA. Retrieved February 7, 2024, from https://www.iea.org/reports/global-ev-outlook-2023

Jonathan, Thomas (2016) *Driving electric vehicles at highway speeds. The effect of higher driving speeds on energy consumption and driving range for electric vehicles in Australia.* https://www.sciencedirect.com/science/article/pii/S1364032116301721

König, A., Nicoletti, L., Schröder, D., Wolff, S., Waclaw, A., & Lienkamp, M. (2021). An Overview of Parameter and Cost for Battery Electric Vehicles. World Electr Veh J, 12(1), 21. https://doi.org/10.3390/

Kuby M, Lim S (2005) The flow-refueling location problem for alternative-fuel vehicles. Socioecon Plann Sci 39(2):125–145. https://doi.org/10.1016/j.seps.2004.03.001

Lee YG, Kim HS, Kho SY, Lee C (2014) User equilibrium–based location model of rapid charging stations for electric vehicles with batteries that have different states of charge. Trans Res Record J Trans Res Board 2454(1):97–106. https://doi.org/10.3141/2454-13

Masoumi H (2020) Urban commute travel distances in Tehran, Istanbul, and Cairo: weighted Least Square models. Urban Sci 4(3):39. https://doi.org/10.3390/urbansci4030039

MirHassani SA, Ebrazi R (2013) A flexible reformulation of the refueling station location problem. Transplant Sci 47(4):617–628. https://doi.org/10.1287/trsc.1120.0430

NTV (2023, November) Türkiye'de satılan elektrikli otomobiller ve güncel fiyatları. Retrieved February 7, 2024, from https://www.ntv.com.tr/galeri/otomobil/turkiyede-satilan-elektrikli-otomobiller-ve-guncel-fiyatlari,uzO8fADG-UGxFg5I0H7GEQ

OpenStreetMap contributors (2024). OpenStreetMap [Data set]. OpenStreetMap Foundation. Available as open data under the Open Data Commons Open Database License (ODbL) at openstreetmap.org, (Accessed on March 18th, 2024)

Rauh N, Franke T, Krems JF (2014) Understanding the impact of electric vehicle driving experience on range anxiety. Hum Factors 57(1):177–187. https://doi.org/10.1177/0018720814546372

Seker (2015, June) Çizge Teorisi (graph theory). YBS Ansiklopedi. Retrieved February 7, 2024, from https://ybsansiklopedi.com/wp-content/uploads/2015/05/cizge_teorisi.pdf

Traffic and Transportation Information for 2022: Annual Average Daily Traffic Values and Transportation Information of Highways and State Roads by Traffic Zones (2023). Karayolları Genel Müdürlüğü. Retrieved February 7, 2024, from https://www.kgm.gov.tr/SiteCollectionDocuments/KGMdocuments/Istatistikler/TrafikveUlasimBilgileri/22TrafikUlasimBilgileri.pdf

The UN Climate Change Conference in Glasgow (2022) COP26 declaration on accelerating the transition to 100% zero emission cars and vans. Retrieved February 7, 2024, from https://www.gov.uk/government/publications/cop26-declaration-zero-emission-cars-and-vans/cop26-declaration-on-accelerating-the-transition-to-100-zero-emission-cars-and-vans

Wang YW, Lin CC (2009) Locating road-vehicle refueling stations. Trans Res E Logistics Trans Rev 45(5):821–829. https://doi.org/10.1016/j.tre.2009.03.002

Worldwide Harmonised Light Vehicle Test Procedure (2021) VCA. Retrieved February 7, 2024, from https://www.vehicle-certification-agency.gov.uk/fuel-consumption-co2/the-worldwide-harmonised-light-vehicle-test-procedure/

Xiang Y, Zhou H, Yang W, Liu J, Niu Y, Guo J (2017) Scale evolution of electric vehicles: a system dynamics approach. IEEE Access 5:8859–8868. https://doi.org/10.1109/access.2017.2699318

Gamze Neşvat, born in 1994 in Bursa, Turkey, completed her high school education at Bursa Kız Lisesi. She earned a Bachelor of Science degree in Industrial Engineering from Yalova University in 2019. Upon graduation, she began her career as a Research & Development Engineer at Üçge Mağaza Ekipmanları A.Ş. in Bursa, where she gained valuable experience in process optimization. She then transitioned to the role of Business Development & Performance Management Specialist at Merter Helva A.Ş. (Koska) in Istanbul. Concurrently, she simultaneously pursued and earned her Master's in Business Administration from Istanbul University in 2024. In her role at Koska, she spearheaded initiatives to enhance operational efficiency, optimize CRM usage, and develop key performance indicators. Gamze is proficient in various data analytics tools, including SQL, Python, R, Tableau, and BigQuery.

Eyüp Çetin is a tenured Professor of Operations Research and Statistics in the Department of Quantitative Methods at the School of Business, Istanbul University, and an Affiliate Distinguished Professor at York University, Canada. He is also affiliated with New York Business Global, USA. He has held various managerial positions. He is former and founding Dean of the Faculty of Transportation and Logistics and former Vice Dean of the School of Business at Istanbul University. He received a Ph.D. in quantitative methods from Istanbul University in 2004. He has taught courses in management science, statistics, business analytics, stochastic models, game theory and health analytics to all levels of business, engineering, healthcare management students: undergraduates, MBAs and Ph.D. students. He has been a consultant for numerous companies, and he has taught statistical & decision modelling and business analytics in even global companies such as Vodafone. His current research focuses on business analytics, data science, digital business & health, healthcare management science, mathematical medicine, operations research, mathematical & statistical modelling. Dr. Çetin is also Founder and Editor-in-Chief of prestigious international academic journals such as European Journal of Pure & Applied Mathematics.

Part III
New Approach in Auditing, Finance and Management

Chapter 5
Risks in the Creation of the Internal Control System of Digital Transformation and Internal Audit Function in Management of the Risks: Risk Awareness Investigation in Turkiye

Seval Selimoğlu and Mehtap Altunel

Abstract This study aims to conduct awareness research in Turkey on how the internal audit function follows/will follow to manage the risks encountered in the digital transformation process and to present suggestions as a result of this research. In the doctoral thesis, the two-phase Delphi technique was used. In the first phase, semi-structured interviews were conducted with 13 participants: academics, internal auditors working in public, independent auditing institutions, and law-making institutions. This stage constitutes the qualitative part of the study. As a result of the interviews, in the second stage, the survey form consisting of 65 statements was presented to the participants and their opinions were taken. The second and third phases constitute the quantitative part of the study. In this study, semi-structured interviews, which are the first phase of the study, are presented. As a result of the study, it was found that the participants were aware of the effects of digital transformation on internal audit and that the institutions had expectations from the internal audit function within the framework of the consultancy role in the digital transformation process.

Keywords Risk management · Internal control system · Digital transformation · Management risk

S. Selimoğlu (✉)
Faculty of Economics and Administrative Sciences, Business Administration Department, Anadolu University, Eskişehir, Turkey
e-mail: sselimoglu@anadolu.edu.tr

M. Altunel
Eskişehir, Turkey

K. T. Çalıyurt (ed.), *New Approaches to CSR, Sustainability and Accountability, Volume VI*, Accounting, Finance, Sustainability, Governance & Fraud: Theory and Application, https://doi.org/10.1007/978-981-96-5937-1_5

5.1 Introduction

Considering that the digital transformation effect experienced with the fourth industrial revolution will continue day by day, it indicates that the change in the business environment of institutions will continue. Digital technologies have necessitated changes in many issues, from the behavior of individuals to the way institutions do business (Verhoefa et al. 2021, p. 890). Although digitalization in the business environment offers institutions competitive advantages, effectiveness, and efficiency, it also brings with it some risks. The literature emphasizes the importance of internal audit focusing on strategic risks by adding value by providing more active support to risk management at different levels and acting in line with the strategic objectives of the institution (Allegrini and D'Onza 2003, p. 199). On the other hand, while Industry 4.0 technologies require changes in the activities carried out by the internal audit function, they raise questions about the rapid growth of digitalization and the role of the internal audit function (Betti et al. 2021, p. 873). It has been found that digital transformation affects three areas in internal auditing. The first is that it expands the scope of auditing, the second is the increase in the need for consultancy within the framework of the role of internal auditing, and the third is the increase in auditor competence and use of technology needed within the scope of internal auditing working practices (Betti and Sarens 2021). In order to contribute to the literature on this field, discussions were held and suggestions were made on how the internal audit function contributes to the risks created by the digital transformation process, what characteristics the internal auditor should have in order to manage the process correctly, and legal regulations.

In the study, first of all, its effects within the framework of digital transformation and internal audit will be discussed. This study is derived from the doctoral thesis. Delphi technique was used in the thesis, and this technique is a two-phase technique. In the first phase, semi-structured interviews were conducted with 13 participants, consisting of academics, internal auditors working in public, independent auditing institutions, and law-making institutions. This phase constitutes the qualitative part of the study. As a result of the interviews, in the second phase, the survey form consisting of 65 statements was presented to the participants and their opinions were taken. The second and third phases constitute the quantitative part of the study. In this study, semi-structured interviews, which are the first phase of the study, are presented. In the study the literature, method, and findings of the study and the conclusion are included.

5.2 Developments in the Field of Internal Audit During the Digital Transformation Process

According to international internal audit standards 2120.A1, "The internal audit activity must evaluate risks related to the organization's governance, operations and information systems in relation to the organization's achievement of its strategic objectives." The statement explains the role of internal audit in providing assurance regarding the risks the institution may encounter in its strategies (IIA 2017, p. 16). Therefore, internal audit also plays a major role in digital transformation, which has an important place in the strategies of businesses. Within the scope of consultancy duty, which is one of the services of internal audit, it is necessary to inform the board of directors and executive management levels about digital transformation and raise awareness. In addition, they are responsible for providing the contribution of senior management and all units in making digital transformation plans. It is of great importance for internal audit to improve itself in this field in order to fulfill its duty effectively in the digital transformation process (Kaya 2018). How internal auditing affects the changing business environment as a result of digital transformation is summarized by Betti and Sarens (2021, p. 210) with the model in Fig. 5.1.

Within the scope of this study, it was found that digitalization affects the internal audit function in three areas. These are: the scope of the internal audit function, the role of internal audit, and the working practices of the internal audit function. The change in the scope of digitalization internal audit causes businesses to face new risks as they add digital to their strategies in order to gain competitive advantage. These risks may be internal risks arising from the disruption of the technology used, or they may originate from outside the institution in the form of cyber risks. As a result, the internal audit function is not limited to the institution but its scope has expanded. In addition, today's business environment has variable, uncertain, complex, and ambiguous (Volatility, Uncertainty, Complexity, Ambiguity (VUCA) characteristics (Schoemaker et al. 2018, p. 15), and the audit plan needs to be made more effective in order to respond to newly emerging risks. It needs to be agile. Second, it has been concluded that with digitalization, the role of internal audit will also change. At this point, the senior management needs more consultancy activities in facing the difficulties that have been experienced or are likely to be experienced with the technology used, rather than the role of a controller who only adopts assurance activities with digital transformation in the internal audit function. At the same time, the internal audit function should not take a role in the definition of strategies, but the increasing need for consultancy activities of institutions due to digitalization causes internal audit to have an indirect impact on business strategies. As a result of the increasing demand for the consultancy activity of internal auditing, there is a divide between independence and the added value created by the consultancy activity. Third, digitalization has affected the working practices of the internal audit function. At this stage, digitalization will influence the use of tools that will make internal audit work practical. Since the technology used allows testing all data sets, it will ensure that reporting is significantly accurate. It will also reduce the time internal auditors spend on analysis. Internal auditors' use of new technologies will enable in-depth and advanced predictive analysis of how anomalies will affect the organization in the future. On the other hand, the use of technology in internal auditing creates both installation costs and the cost of acquiring the necessary skills for internal auditors. Because with digital transformation, digital skills have become important for internal audit departments (Betti and Sarens 2021, pp. 202–208). In summary, digital transformation has both expanded the risk areas of institutions and

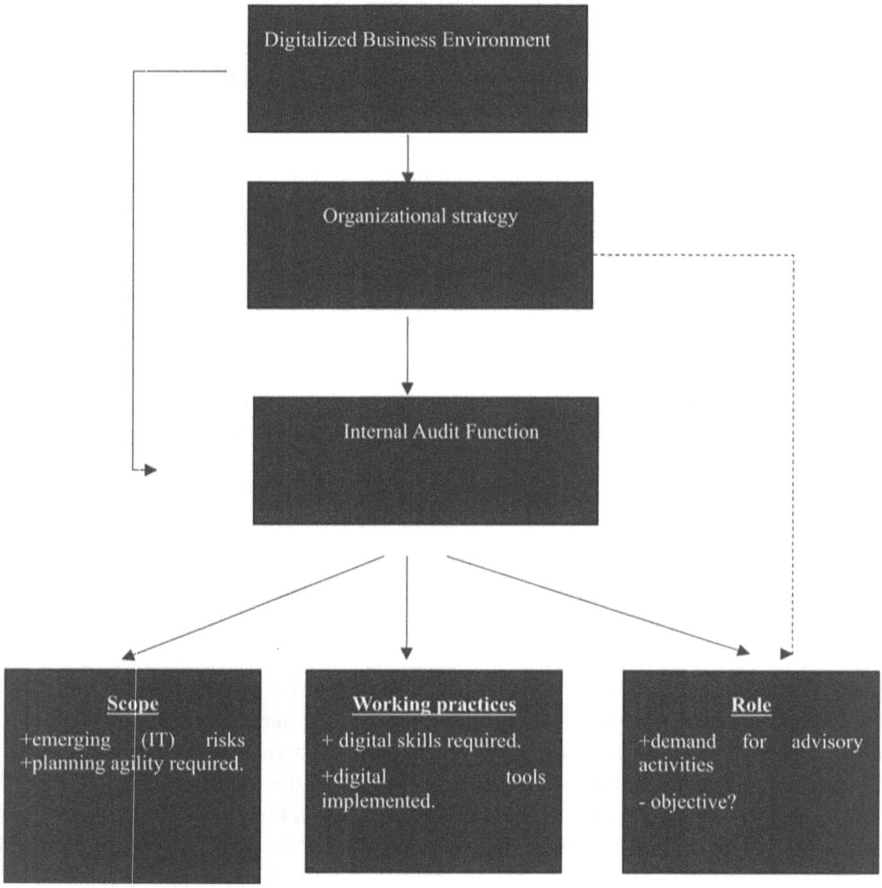

Fig. 5.1 The impact of the digital business environment on the internal audit function (Betti and Sarens 2021, p. 210)

increased the need for consultancy activities regarding these risk areas. In addition, developments in technology will contribute to the more effective conduct of internal audit activities.

5.3 Literature

The study conducted by Akçakanat et al. (2021) reveals the cyber risks of institutions and provides information about managing these risks. In this context, the information and information systems presented within the framework of cyber security in the activity and integrated reports of the top ten banks according to asset size were examined. As a result of the study, it was found that the banks have a structure that complies with national and international regulations, 120 internal audit departments carry out relevant audit activities, relevant training is provided in this context, and technology is followed for data security.

Bircan's (2020) study aimed to measure the awareness of internal auditors within the framework of the structural change and transformation process in internal auditing. Within the framework of this study, data was collected from 51 internal auditors by survey method. As a result of the study, it was found that internal auditors have mental awareness of change and transformation, but there are deficiencies in the internal audit practices phase.

In the studies of Güler and Arkın (2019), the role of internal audit was evaluated within the framework of cyber security and cyber hygiene. It has been discussed that internal audit has a strong potential in ensuring cyber security and that the cyber responsibility of internal audit increases day by day. In the studies of Selimoğlu and Altunel (2019), the role of internal audit in the protection phase from cyber risks was discussed. The IIA emphasizes the triple line of defense, which provides a systematic perspective on ensuring cyber security.

Betti and Sarens (2021) conducted research on how internal audit has developed in the digitalizing business environment. Interview technique, which is a qualitative research method, was used. Twenty-nine people were interviewed, including the audit committee member, the internal audit manager, and the internal auditor. As a result of the study, it was found that the digitalized business environment affects the internal audit function in three aspects. The first is that the scope of internal audit has expanded. It was emphasized that the agility of internal audit planning and the increase in the required digital knowledge and information technology risks, especially cyber security threats, are gaining importance. Second, the demand for consulting activities performed by internal auditors is higher. Third, digitalization has changed the working practices of internal auditors in their daily duties, and new technologies such as data analytics tools are being gradually implemented in internal audit departments and digital skills are considered a critical asset.

In their research, Betti et al. (2021) aim to investigate how the internal audit function changes its activities and practices regarding the digitalization of institutions. This research specifically examined the use of data analytics and the performance of consulting activities by internal auditors. A survey technique was applied to internal auditors at the Internal Auditors Institute in the United States and a response was received from 82 participants. As a result of the study, it was found that the digitalization of the business environment is positively related to the use of data analytics for internal auditors, and the level of digitalization of institutions due to the use of data analytics technologies by the internal audit function positively affects the performance of consulting activities.

In the studies of Selimoğlu and Saldı (2022), the position of the internal audit activity in cyber security governance was determined. Opinions were collected from experts using the Delphi technique. The study carried out in the banking sector reached conclusions on the headings of "Responsibility", "Ethical Principles", "Competence", "Confidentiality, Integrity and Accessibility", "Governance",

"Corporate Governance 122 Principles", "Legal and International Policy Framework".

The study conducted by Güngör (2021) has three purposes; first, to detect the cyber security activities of publicly traded companies in Turkey; The second is to determine the cyber security activity effectiveness of internal audit activities and to determine whether this effectiveness differs according to businesses. Third, to determine whether there is a relationship between the cyber security activity level of enterprises and the cyber security activity effectiveness of internal audit. In line with the purpose of the study, a survey was applied to 222 internal auditors working in public companies listed on Borsa Istanbul. As a result of the study, it was found that cyber security levels differ depending on the type of businesses, the cyber security activity effectiveness of internal audit varies according to business types, and there is a strong relationship between the cyber security activity level and the cyber security activity effectiveness of the internal audit unit.

Lois et al. (2021) aimed to examine the variables affecting cyber security and related internal audit in their research. For this purpose, data was collected by applying a survey technique to internal auditors in companies traded on the Athens Stock Exchange. As a result of the study, key factors affecting cybersecurity were identified, including the degree and nature of collaboration between IT personnel and auditors and training in information technology. This research highlights the need for audit and auditors to expand their knowledge of cybersecurity in order to achieve success in digitalized auditing.

In their study, Tiberius and Hirth (2019) examined the changes that German audit professionals expect in audit practices in the next five to ten years. In the study, data was collected using the Delphi technique. The study touches upon the perception of auditing, the auditor-client relationship, regulations, structural and procedural changes for audit firms, and the profile of the auditing profession. As a result of the study, it was concluded that extensive changes are not expected within the determined time period and that the annual audit will gradually evolve into a continuous audit approach. Despite the predominantly vague views, experts have determined that new technologies will not replace the auditor but will instead provide assistance and support, and although the demands of the job may make it difficult to remain in the profession, disruptive effects on auditors' workplaces are not expected in the near future.

5.4 Method

In the doctoral thesis, data were collected using the Delphi technique. The Delphi technique is a method for structuring a group communication process. Thus, the process is effective in gaining consensus among a group of individuals with expertise on a particular topic where the information sought is subjective and participants are separated by physical distance (Khayun et al. 2012, p. 34). Delphi technique is defined by Macmillan (1971, p. 1) as a method of revealing and improving group judgments. The method requires knowledgeable and expert participants to answer questions individually and transmit the results to a central researcher. The researcher identifies contributions by looking for central and extreme tendencies and their justifications, and the results are returned to the respondents. Respondents are then asked to represent their views with the help of the input provided by the researcher. This process continues until the researcher sees that a consensus has emerged. The technique aims to eliminate bias that occurs when different groups of experts come together. In the Delphi technique, experts do not know who the other experts are throughout the process (Grisham 2009, p. 114).

In this study, semi-structured interviews, which are the first phase of the study, are presented. The first question is about getting to know the participant profile. Table 5.1 shows a participant profile.

Q1—What is your role in the organization and how long have you been working in this position?

5.5 Findings

Q2—What do the Fourth Industrial Revolution, Fourth Industrial Revolution technologies, digital transformation remind you of? What are the reflections of digital transformation on your institution?

P3—"The fourth industrial revolution, especially with 5G technology, calls for the ability to remotely control the machines we use in our business and private lives by connecting to the internet, and the development of machine and robotic technologies, especially the reduction of human labor for routine work, and the greater integration of machines and robots into production processes. As an audit and consultancy firm, the reflections of digital transformation on our institution have been more indirect compared to real sectors such as production, service and retail."

P7—"It makes a few senses. Firstly, it makes sense for automation to accelerate with digital transformation and for machines to take part in business life as digital workforce. There is a process in which IoT systems talk to each other and are implemented as a support mechanism for different decision makers, and this is constantly evolving. These studies have become increasingly common, especially in the manufacturing sector and the automotive industry. Of course, there is a need for an integrated system infrastructure, an integrated system solution, and for us and the consultancy side, the installation of these systems and the processing and tracking of the data to be received from these systems, and the third is the security of these systems; It is security in the sense of cybersecurity. Because, of course, cyber security services are critical and important for us here in order to prevent any negative situation, such as outages in power lines, or interruptions in the production line."

P13—"I think the digitalization of institutions means that most machines talk to each other and work integrated with data. Of course, the reflections of this on the institution are to automate the systems as much as possible, to reduce human intervention and time spent as much as possible, and to be more effective, easier,

Table 5.1 Characteristics of the participants in the study

Code	Title	Gender	Educational level	Occupational experience (year)	Organization
Academicians					
P1	Assoc. Prof.	Male	PhD	20 years and above	Public University
P2	Assoc. Prof.	Female	PhD	15–19	Public University
P5	Assoc. Prof.	Female	PhD	20 years and above	Foundation University
P9	Assoc. Prof.	Male	PhD	15–19	Foundation University
Internal auditor					
P3	Internal auditor	Male	Master	10–14	International Audit Company
P4	Internal auditor	Female	Bachelor	5–9	Big Four
P7	Partner/internal auditor	Male	–[a]	–	Big Four
P8	Internal auditor	Female	–	–	Big Four
P13	Internal auditor	Male	Master	15–19	Big Four
Public internal auditor					
P6	Public internal auditor	Male	Bachelor	20 years and above	Public Organization
P10	Public internal auditor	Male	PhD	20 years and above	Public Organization
Professional association					
P11	Internal auditor	Male	Master	20 years and above	ISACA
P12	Internal auditor	Male	Master	15–19	ISACA

[a]The information in this field is demographic information and is created in line with the results obtained from the Delphi second round survey. Since K7 and K8 did not participate in the Delphi second round survey, information about their educational background and professional experience could not be obtained

accessible from anywhere in the world, and workable from anywhere in the world as long as you are connected to the internet. With this pandemic, we have seen once again how important it is to create systems.... What challenges companies like us are customer habits. No matter how much digital transformation you achieve, if the customers you work with are not compatible with it, it remains a bit out of date."

Q3—Considering institutions, what do you see as the main risks that the digital transformation will bring?

P2—"I think the most important risks that will be experienced with digital transformation are the protection of personal data and cyber security risks."

P3—"Significant changes in the way of doing business and the increase in machine and robotic processes instead of human labor will end some business lines but will create new business lines. For this reason, unemployment seems to be an important problem for people and societies that cannot integrate into these processes and renew themselves...."

P4—"The biggest risk from digital transformation is in the field of data. Here, it should not be considered just as data loss; it is a great risk that the data cannot be kept in a holistic and meaningful way, that is, the data cannot be interpreted correctly.… The second risk is cyber security risk. It has become more prominent especially with the remote working process."

P7—"I care about cyber security risk. One of the biggest risks is that the data produced may leak out, that data or systems may be attacked from outside, and that the system may be interrupted for a short or long time. Secondly, digital workforce may come to mind. The process to increase productivity should proceed in this way…"

Q4—What is the role of the internal audit function against the risks encountered with digital transformation?

P8—"Internal audit must first inform the institution about the risk. In fact, I think internal auditing is not only an auditor but also a consultancy hat. The people who know the job best are the process owners. But the internal auditor can very well identify risks, guide the organization in this regard, and offer action plans to eliminate or minimize the risks. Therefore, I can summarize it as he should use both hats."

P12—"Internal audit functions must first conduct in-depth analysis on the opportunities and threats brought by new technologies and guide the senior management well on these issues. The consultancy role of internal audit units, which are generally evaluated through the assurance function, is very valuable in this respect. In addition, it is very critical to reveal invisible or not initially understood problems through assurance activities and to attract the attention of relevant stakeholders on these issues. For example, a company may want to take advantage of cloud services. This is not an issue that can be solved in two days by paying money. It requires a total cultural change. It also carries very significant risks. Simply put, there is a data migration problem. I'm sure many senior managers are not aware of the vital critical points regarding data migration. This should be one of the most important topics of internal audit. It should be ensured that new technologies are well analyzed and understood."

Q5—How would you evaluate the assurance and consultancy role of internal audit within the framework of digital transformation?

P5—"Initially, consultancy can make a significant contribution. As the level of institutional maturity increases, assurance activities can be included."

P7—"… Since we will express our opinions on both the development of competencies and the improvement of processes, one of our main functions is to provide remedial suggestions in addition to identifying deficiencies in the current processes. Therefore, the need for a consultancy role is high…"

P9—"…If there is internal audit in a company, there is no such thing as 100% risk elimination or assurance. But a significant part of it is resolved. The issue of internal audit not being able to provide 100% assurance within the framework of digital transformation will continue. In fact, the situation of not being able to provide assurance may even increase. But on the other hand, the consultancy role will be more prominent. In fact, I think the consultancy role will become more critical."

Q6—How would you evaluate internal audit, which is in the third line role of the triple line model, within the framework of digital risks?

P2—"In the revised triple line model, the importance of internal audit has been brought to the fore and the expectations from internal audit have increased. In this case, in the new changing and digital world, risks arising from digitalization make internal audit activity even more necessary."

Q7—What changes should institutions make in their current internal control system in the face of risks arising from digital transformation? What is the role of internal audit at this point?

P1—"Considering digital transformation as a process of adopting or adapting to some technological innovations means drawing a narrow framework. Digital transformation brings with it a restructuring process that puts technology at its center. Therefore, the change in the institutional structure, the development and/or change of the institution's human resources, and some changes in the corporate culture are part of this transformation..."

P2—"The risk environment and control activities components of the internal control system should be revised to include digital risks. Internal auditing, as a mechanism that ensures the monitoring activity among the five components of the internal control system, can contribute especially with its assurance role in identifying and eliminating the deficiencies of the internal control system. "It can provide consultancy in identifying risks and determining the control activities to be created regarding these identified risks."

P5—"Digitalization should start with a strategy. Within the framework of risk management, necessary control points should be established in accordance with corporate governance principles. Internal audit should assume a strategic role. "It would be appropriate to highlight the consultancy role."

P11—"If a company has started to take digitalization steps in its business processes and has made this a company culture, it must naturally update its internal control structure with this perspective. It is not possible for employees to turn their backs on these innovations while doing their jobs and managers updating their goals and workflows.

Q8—What do you think are the new features that internal auditors should have with the digital transformation process? What should be the focus of internal auditors in this process?

P5—"It is important to quickly establish the necessary infrastructure for internal auditors to have digital literacy as well as financial literacy. Otherwise, it may cause loss of reputation. "An auditor who does not know the business processes he will audit cannot be expected to propose solutions to the issues, develop recommendations and produce value-added work."

P7—"Cyber security and data analysis are important in this sense. Algorithms and data coding will gradually become something that internal auditors can do, too....there needs to be competencies to collect and analyze data and at the same time ensure the security of this data. In fact, you need to know the traditional methods and add technology to them. One of the questions that comes up, especially when we attend meetings with universities, is: Do we need to graduate specifically from the engineering side? No, actually there is no such thing, people from the finance group also need to improve themselves in this field."

P13—"…The internal auditor should check whether integration with the new order has been achieved with digital transformation. It should focus on how tricks can be done in the digital environment. Because in his world until now, he was always focusing on manual tricks that could be done, what could be changed, what could be overcome…."

Q9—What training/certificates should the auditor receive in order to manage the process correctly? What should be done to train internal auditors with the competence required by this era of transformation?

P7—"CIA certificate is already one of the 'must' certificates for us. If we are to manage technological risk, ISO is always risk-based. However, if we are going to engage in studies such as data analysis, we are not looking for a specific certification here, but this could be a plus for us. Must be able to write an SQR-based code and use any data analysis tool while analyzing data. When I think of it as a 'must,' CIA; Again, although CISA is not a priority, it is one of the 'must' ones. We are gradually getting our own teams certified within the framework of robotic process automation (RPA) data analysis… Educational institutions need to design processes to carry out work in the field for the future. In other words, students may be learning about finance, purchasing, financial affairs, production and some business operations, but especially for internal audit activities, what kind of risky areas might be in those processes and what kind of things are encountered in them. Examples of this are really life-related abuse, etc. *'They will improve themselves by dealing with the issues through real-life cases.'"*

P12—"Certifications such as CISA, CRISC, CDPSE, CET offered by ISACA are important for internal auditors in terms of meeting the requirements in the field of technology. 'I also really like the CIA content offered by The IIA.'"

Q10—What main regulations/guidelines should be used when conducting internal audit activities against the risks that come with digital transformation? How would you evaluate the content of these guides?

P1—"The Information and Communication Security Audit Guide published by the Presidency Digital Transformation Office in October 2021 can be used. The guide has sufficient content for the implementation of the audit."

P2—"Information and Communication Security Audit Guide was published in 2021 by the Presidency Digital Transformation Office for public institutions. Since this guide has been prepared considering many standards, it is one of the most important and comprehensive reference sources for public internal auditors and also for businesses providing critical infrastructure services. 'In addition, international standards and frameworks related to information and communication security such as COBIT, ISO 27000, NIST can also be used.'"

P11—"First of all, it would be more beneficial to follow organizations or institutions that can adapt to rapidly changing and developing digital processes and adopt their approaches (certificate, regulation, framework, etc.). A framework that has not been updated for 6-7 years or a certificate content that has not changed for 5-6 years does not contribute much anymore…".

Q11—How would you evaluate the regulations made to manage risks arising from digital transformation in Turkey?

P6—"There was awareness before, but there was no regulation. It was great to have such a regulation (Information and Communication Security Audit Guide) that concerns the entire public. They force institutions to obtain information security certificates. The Information and Communication Security Audit Guide issued by the Presidential Digital Transformation Office was very important because it required all public institutions. Again, Law No. 5018 is important because it refers to the internal control system."

P10—"Information and Communication Security Audit Guide by the Presidential Digital Transformation Office is sufficient as documentation…".

5.6 Conclusion

Digital transformation has changed the way organizations do business. Cost, time, etc., provide advantages in terms of subjects. On the other hand, it brings with it many risks. Cyber security risk is shown as the main risk by the IIA. In this study, it has been revealed with brief theoretical explanations that the change in internal audit activities and internal auditor competency is inevitable in this process where there is a change in the ways of doing business, regardless of private or public institutions. In addition, the role of internal audit in order to manage the process correctly has been tried to be explained. Subsequently, the study conducted research on the main risks seen by internal auditors working in academics, public institutions, independent audit institutions, and professional organizations in the digital transformation process, the role of the internal audit function in managing this process correctly, the competence of the internal auditor, and legal regulations.

As a result of the semi-structured interviews, which were the first round of the thesis study in which data was collected using the Delphi technique, the following conclusions were reached:

- All participants are aware of Industry 4.0, digital transformation, and the impact of digital transformation on the internal audit function. There are differences in terms of the reflection of technologies on institutions. It is seen that Industry 4.0 technologies are more concentrated in private institutions.
- When digital transformation and internal audit are considered together, it has been reached that the first mentioned issue is the concept of real-time monitoring.
- It has been found that digitalization causes institutions to encounter new risks and the size of the risk to change. It has been found that ensuring data integrity, protecting personal data, information security, and cyber security risks come to the fore. Other important risks include the fact that some business lines are in danger of disappearing due to the decrease in the need for manpower due to technological developments or the current competence not meeting expectations.

- It is emphasized that the pandemic is effective in bringing the cyber security risk to the fore.
- Institutional managers' resistance to technology is seen as one of the limitations faced by internal auditors.
- Within the framework of digital transformation, it has been found that the consultancy role of the internal audit function has come to the fore and the need for a consultancy role has increased.
- It has been concluded that the independence and impartiality of auditors will not be negatively affected by the consultancy role expected from internal auditors in the digital age. Compliance with the triple line model is seen as an effective factor in ensuring that independence and impartiality are not damaged.
- It is important that the digital transformation process should be among the strategic goals and progress in a way that covers the entire corporate culture, from the corporate structure to human resources. Considering that digital transformation in institutions is a process beyond the acquisition of a technology and its adaptation to the institution, a strategy must be determined to cover the entire institution. Additionally, the internal audit function should take part in this process in a consultancy role.
- It is important that internal auditors need to improve themselves during the digital transformation process and that people have cross-functional functions.
- Internal auditing for information security has become mandatory in public institutions, and one of the audit guide requirements is the obligation of a CISA certified internal auditor. However, it appears that public institutions do not have competent personnel to meet this requirement.
- Within the framework of legal regulations, it is seen that BRSA and CMB institutions are ahead of the practices in other public institutions with their regulations and practices regarding information security.

As a result of the study, the main work should be to explain the change experienced by internal auditors to the corporate managers, in creating awareness and turning digital transformation into a corporate culture. On the other hand, since the main emphasis is on the competence of the internal auditor, it will be beneficial to include people who are skilled in digital technologies in the internal audit teams in terms of contribution to the institution. From the perspective of public institutions in Turkey, although guidance has been published to manage risks arising from digital transformation, personnel competence is not at the desired level. It is important to recruit competent personnel to institutions as soon as possible in order to manage the process effectively. Another important issue is educational institutions. It is important to draw attention to the internal audit profession, which is important for educational institutions and businesses. Following a path by considering both control and technology issues together in this process will be beneficial for businesses to reach the capabilities they need.

Acknowledgment This thesis study was supported within the scope of the project no. 2108E221 accepted by Anadolu University Scientific Research Project Commission. This research was conducted by Mehtap Altunel at Anadolu University Social Sciences Institute, Department of Business Administration (Accounting) titled "Risks in the Creation of the Internal Control System of Digital Transformation and Internal Audit Function in Management of the Risks: Risk Awareness Investigation In Turkey" derived from his doctoral thesis.

References

Akçakanat Ö, Özdemir O, Mazak M (2021) İşletmelerde Siber Güvenlik Riskleri ve Bilgi Teknolojileri Denetimi: Bankaların Siber Güvenlik Uygulamalarının İncelenmesi. Mehmet Akif Ersoy Üniversitesi Uygulamalı Bilimler Dergisi 5(2):246–270

Allegrini M, D'Onza G (2003) Internal auditing and risk assessment in large Italian companies: an empirical survey. Int J Audit 7:1191–1208

Betti N, Sarens G (2021) Understanding the internal audit function in a digitalised business environment. J Account Org Change 17(2):197–216

Betti N, Sarens G, Poncin I (2021) Effects of digitalisation of organisations on internal audit activities and practices. Manag Audit J 36(6):872–888

Bircan NG (2020) İç Denetimde Yapısal Değişim ve Dönüşüm: İç Denetçilerin Farkındalığı ve Beklentileri Üzerine Bir Araştırma. Muhasebe Enstitüsü Dergisi 63:67–83

Grisham T (2009) The Delphi technique: a method for testing complex and multifaceted topics. Int J Manag Proj Bus 2(1):112–130

Güler A, Arkın AK (2019) Siber Hijyenin Sağlanmasında İç Denetimin Rolü. Denetişim Dergisi 9(19):17–40

Güngör N (2021) İç Denetimde Bilgi Teknolojileri Ve Siber Güvenlik: Borsa İstanbul Şirketlerinde Bir İnceleme. İstanbul Üniversitesi, Sosyal Bilimler Enstitüsü, İstanbul

IIA (2017) Mesleki Uygulama Çerçevesi Kapsamında Uluslararası İç Denetim Standartları (Standartlar). The Institute of Internal Auditors. https://www.theiia.org/globalassets/documents/standards/standards-2017/ippfstandards-2017-turkish.pdf. Accessed 15 Feb 2021

Kaya B (2018) Dijital Dönüşümde İç Denetçilerin Rol ve Sorumlulukları. https://bertankaya.net/2018/06/dijital-donusumde-ic-denetcilerin-rol-vesorumluluklari/. Accessed 9 Feb 2021

Khayun V, Ractham P, Firpo D (2012) Assessing E-excise success with Delone and McLean's model. J Comput Inf Syst 52(3):31–40

Lois P, Drogalas G, Karagiorgos A, Thrassou A, Vrontis D (2021) Internal auditing and cyber security: audit role and procedural contribution. Int J Manag Financ Account 13(1):25–46

Macmillan TT (1971) The Delphi technique. In: Paper presented at the annual meeting of the California Junior Colleges Associations Committee on Research and Development, pp 1–24

Schoemaker PJ, Heaton S, Teece D (2018) Innovation, dynamic capabilities, and leadership. Calif Manag Rev 61(1):15–42

Selimoğlu SK, Saldı HM (2022) Türk Bankacılık Sektöründe İç Denetim Yoluyla Siber Güvenlik Yönetişimi. İşletme Akademisi Dergisi 3(2):161–187

Selimoğlu S, Altunel M (2019) Siber Güvenlik Risklerinden Korunmada Köprü ve Katalizör Olarak İç Denetim. Denetişim Dergisi 19:5–16

Tiberius V, Hirth S (2019) Impacts of digitization on auditing: a Delphi study for Germany. J Int Account Audit Tax 37:1–14

Verhoefa PC, Broekhuizen T, Bart Y, Bhattacharya A, Dong JQ, Fabian N, Haenlein M (2021) Digital transformation: a multidisciplinary reflection and research agenda. J Bus Res 122:889–901

Seval Selimoğlu was born in Eskişehir in 1962. After completing her primary and secondary education in Eskişehir, she graduated from Anadolu University, the School of Economics and Administrative Sciences, Faculty of Business in 1984. She started his academic career in 1986 as a research associate in Anadolu University, the School of Economics and Administrative Sciences, Faculty of Business, Department of Accounting and Finance. She completed her master's degree at Anadolu University, Institute of Social Sciences, Department of Accounting in 1987. She obtained an International MBA (IMBA) degree in the United States of America (OHIO-Cleveland) at Baldwin-Wallace University in 1990. She completed her doctorate at Anadolu University, Institute of Social Sciences, Department of Accounting in 1995. Her doctoral dissertation in the audit field was published as a book in 1996 by the Capital Market Board. She received the title of Assistant Professor Doctor in Accounting-Finance in 1995, Associate Professor Doctor in 2002, and Professor in 2007. Her primary academic studies are in International Independent Audit and Assurance Standards, Independent Audit, Ethics, Accounting Fraud, Corruption, Internal audit, Financial Statements Analysis and Accounting. She has lots of books, articles, assertions, and studies on EU projects published nationally and internationally. She served in various commissions established in corporations such as the Union of Chambers of Certified Public Accountants/ Education Department, Public Accounting Oversight Board of Turkey. She also has the titles of Certified Public Accountant (CPA) and Independent Auditor and serves as an expert in courts related to her area of specialization. She is married.

Mehtap Altunel, born in 1989 in Artvin, Turkey, completed her high school education at Artvin Kazım Karabekir Anadolu Lisesi. She earned a Bachelor of School Of Economics and Administrative Sciences from Akdeniz University in 2008. Afterward, she began her Master in Business Administration from Suleyman Demirel University in 2015. She completed her thesis period at Recep Tayyip Erdogan University. In 2018, she completed her master's thesis and earned the doctoral program at Anadolu University. She simultaneously prepared for the CPA exam during his doctorate program and passed the exam. She completed her doctoral thesis in 2023. In 2024, she was accepted as a financial supervisor in a manufacturing company, and she still works.

Chapter 6
Contractor's Bidding Decision Modeling, Calculating Bid Price, and Distributing to Maximize Cash Flow

Cihat Doğan and Eyüp Çetin

Abstract In the construction industry while contractors bid for jobs that are highly profitable and will maximize cash flow in a limited time among many requests for proposals in today's competitive conditions, they experience intuitive and experiential difficulties in making a balanced decision between the benefit of the possibility of winning high-return projects and the cost of bidding. In this study, a decision support model to be used in this decision-making process of a contractor company operating in the electromechanical contracting sector abroad: together with a multi-criteria valuation analysis, one of which determines the total bid price and the other of which distributes this amount to the quantities of work breakdowns in a way that maximizes cash flow. The Analytical Hierarchy Process (AHP) is employed and two linear programming methods are studied.

Keywords Bidding decision · Cash flow · Total price · Contractor · Analytical hierarchy process · AHP · Multi-criteria decision-making

C. Doğan
Graduate School of Business, School of Business, Istanbul University, Avcilar, Istanbul, Turkey
e-mail: cdogan@sasel.com.tr

E. Çetin (✉)
Department of Quantitative Methods, School of Business, Istanbul University, Avcilar, Istanbul, Turkey

Laboratory for Industrial and Applied Mathematics, York University, Toronto, ON, Canada

New York Business Global, College Park, MD, USA
e-mail: eyup.cetin@istanbul.edu.tr

K. T. Çalıyurt (ed.), *New Approaches to CSR, Sustainability and Accountability, Volume VI*, Accounting, Finance, Sustainability, Governance & Fraud: Theory and Application, https://doi.org/10.1007/978-981-96-5937-1_6

List of Abbreviations

AHP	Analytic Hierarchy Process
ANP	Analytical Network Process
ÉLECTRE	ÉLimination Et Choix Traduisant la REalité
MADM	Multi-Attribute Decision-Making
MAUT	Multi-Attribute Utility Theory
MCDM	Multi-Criteria Decision-Making
ORESTE	Organization, Rangement Et Synthese De Donnes Relationnelles
PROMETHEE	Preference ranking organization method for enrichment evaluation
SAR	Saudi Arabian Riyal
TOPSIS	Technique for Order of Preference by Similarity to Ideal Solution
VIKOR	VIseKriterijumska Optimizacija I Kompromisno Resenje

6.1 Introduction

For contractor companies working in the construction sector, tenders are possible business opportunities, and when they are invited to tender, they have to respond to the invitation by quickly deciding whether the tender is worth devoting time and resources (and naturally cost) to them. In the sector, companies generally make these decisions based on experience and managerial intuition along with available data about the tender. In order to minimize the risks arising from today's competitive conditions and time constraints, it has now become a necessity to support these intuitive approaches with scientific methods and to prevent irreversible and highly harmful decisions from being taken. The decision process requires evaluation at different decision levels within the company to reduce costs and increase revenue. This evaluation is affected by many criteria, from the suitability of the company's financial and technical capabilities to the requirements of the job, the economic return of the job, the suitability of the existing business structure, the need for the job, or the strategic experience it will bring to the company.

Different techniques are used to solve problems known as multi-criteria decision-making (MCDM) in the literature. Multi-criteria decision-making is the process of reaching the best possible outcome according to certain rules. Real-life problems often contain the same basic and sometimes conflicting criteria. It is difficult to find a solution that satisfies all the selection criteria. In such problems, an optimum solution is sought within the framework of predetermined rules (Vahdani et al. 2010). In situations where there are too many criteria and too many alternatives, decision-making becomes even more difficult. There are many different criteria decision-making methods used in such cases. The most common of these are: Systems such as AHP, ANP, VIKOR, ELECTRE, PROMETHEE, TOPSIS, Goal Programming, Linear Programming, ORESTE, MAUT and various Artificial Intelligence models (Boran et al. 2009).

In this study, the weights of the decision criteria were determined using the Analytical Hierarchy Process (AHP), one of the MCDM methods, and the tender to be bid on was selected according to these weights. The bid price of the selected tender was calculated as a function of the result obtained from these criteria, and this price was distributed to the business groups in the bid quantity in a way that would yield maximum profit with the most ideal cash flow. Which of the tender selection criteria is more important was determined by taking the opinions of internal decision-makers and experts.

In the second part of the study, the decision to bid for the tender, the decision to determine the bid price, and a brief literature review about AHP are included. In the third part, the method of determining the criteria according to the evaluations of decision-makers and experts and weighting them with AHP, and the method of creating the tender price and distributing it to the quantities, together with the decision to bid/not to bid, are emphasized. In the fourth chapter, an application was made according to the method explained for three tenders specific to the company where the first author works; in the last section, results and findings are shared (Fig. 6.1).

6.2 Related Literature

The number of bids that contractor companies can submit at a given time is limited due to the capacity of the relevant department and cost. Tendering departments have to choose among the invitations for proposals for a wide variety of projects in a short time and to prepare these proposals within a limited time without causing major losses and without damaging the company's reputation. Deciding whether to bid a tender requires preparing well and making a balanced choice between the expected cost of losing the tender and the expected benefit of the project's profit, positive cash flow, and position in the industry.

6.2.1 Bid or No Bid Decision

There are many criteria that affect whether or not the tender will be accepted. These criteria and their importance have been evaluated globally in the construction industry through surveys, expert opinions, and face-to-face interviews in America (Ahmad and Minkarah 1988), England (Shash 2000) and Singapore (Chua and Li 2000) in Saudi Arabia (Bageis and Fortune 2009), in Poland (Leśniak and Plebankiewicz 2015) in Syria (Wanous et al. 2000), and in Nigeria (Oyeyipo et al. 2016). While there are many existing methods for evaluating these multiple criteria (Saaty 1994; Köksalan et al. 2011; Triantaphyllou 2000), many decision support models have been developed in the literature for the comparison of tender invitations (Wanous et al. 2000; Leśniak and Plebankiewicz 2015; Cheng et al. 2011). Jato-Espino et al. (2014) found in his research that 22 different decision models

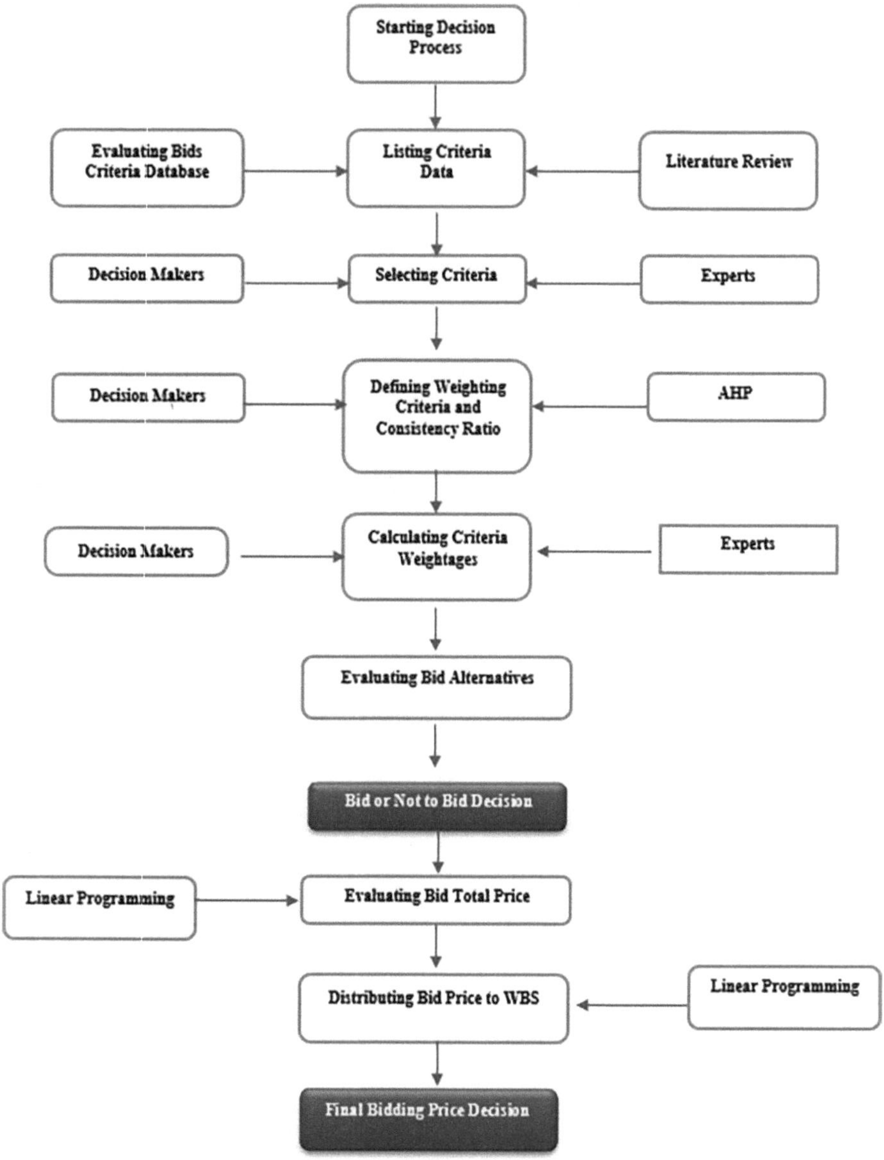

Fig. 6.1 Bid or not to bid and total bid price decision making model

were used in 88 research articles and determined the number of occurrences of these (Table 6.1). According to the research of Ahmad and Minkarah (1988), only 11.1% of construction companies in America used a decision support model, while this rate was 17.6% in England.

Table 6.1 22 different MCDM model per 88 surveys fulfilled in the construction industry (Jato-Espino et al. 2014)

Approach	Method	Number of occurrences
Single	AHP (Analytic hierarchy process)	20
	DEA (Data envelopment analysis)/ELECTRE (Elimination et choix traduisant la realite)	6
	TOPSIS (Technique for order of preference by similarity to ideal solution)	3
	ANP (Analytic Network process)/Delphi/GST (Grey system theory)	2
	Other	1
Hybrid	AHP (Analytic hierarchy process)	26
	FSs (Fuzzy sets)	24
	TOPSIS (Technique for order of preference by similarity to ideal solution)	11
	ANP (Analytic Network process)/MCS (Monte Carlo simulations/ MIVES (Modelo integrado da valor para evaluaciones sostenibles)/ VIKOR (Visekriterijumska optimizacija I kompromisno resenje)	4
	COPRAS (Complex proportional assessment)/GST (Grey system theory)/PROMETHEE (Preference ranking organization method for enrichment of evaluations)/SAW (Simple additive weighting)	2
	Other	1

According to the approaches applied, Bid/Not to bid Models can be classified in three ways (Oo et al. 2007):

- Multiple Attribute Decision-Making models
- Artificial intelligence based models
- Statistical models

6.2.2 Evaluating Total Bidding Price

The price of the bid to be submitted to the tender plays the most important role in getting the job done. In the literature, in addition to models that help determine a rate to maximize the contractor's expected profit while maximizing the probability of getting the job (Mielec et al. 2009; Wang et al. 2007), game theory (Cui and Zhang 2009) and fuzzy set theory (Li et al. 1999) models have also been developed.

6.3 Method

In this section, the methods used in the study and their basic stages are included.

6.3.1 Analytical Hierarchy Process

The Analytical Hierarchy Process (AHP) was developed by Thomas L. Saaty, who passed away in 2017, in the 1970s to solve complex MCDM problems [see some details about AHP in Saaty (2008) and remarks about Saaty's final works in Çetin and Kiremitci (2018)]. AHP determines the priorities of decision-makers, who divide the problems into their components, using their subjective evaluations on the subject, allows the establishment of a hierarchical structure in line with these priorities and can increase the effectiveness of decisions, has more than one sub-goal and a general goal, and the sub-goals are accepted as decision criteria. It is a method in which pairwise comparisons of options are made according to determined decision criteria for the final decision (Fig. 6.2).

In addition to providing convenience and understandability, this method receives criticism on issues such as the validity of changing the order of alternatives, the fact that the evaluation of criterion weights and alternatives is a subjective process, the difficulty of creating pairwise comparison matrices as the number of criteria and alternatives increases, and the results obtained cannot be described as "absolutely correct." The first five stages of the method are as follows: It consists of defining the problem, determining and defining the main and sub-decision criteria, determining the alternatives, creating the hierarchical structure, and determining the relative importance scale.

In the first stage of AHP, the decision problem is determined. In the second stage, an n×n dimensional decision matrix is created based on pairwise comparison of all criteria, including themselves, with each other. All criteria are included in rows and columns so that the comparison can be expressed in the matrix. In other comparisons, numerical values are entered into the matrix based on Saaty's importance table. In the third stage, the percentage importance values of the criteria are determined. In the fourth stage, the consistency in comparing the criteria is measured. In the fifth stage, the percentage importance distributions at m decision points are calculated for each criterion. In the last stage, the result distributions of the alternatives are found.

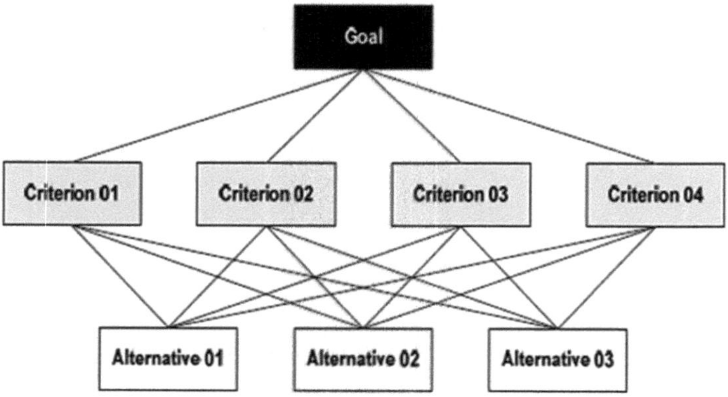

Fig. 6.2 Hierarchy model

6.3.2 Bid or No Bid Decision Model

At this stage, first of all, the overall attractiveness ratio of the job is calculated as "O." Simple additive weighting method is used for this calculation (Eq. 6.1).

Aggregated attractiveness score

$$O = \sum_{i=1}^{n} o_i \cdot w_i, \quad \sum_{i=1}^{n} w_i = 1, \quad o_i \subset 0.1 \tag{6.1}$$

The standard attractiveness score of each criterion, "O," which is the weighted average of the "O_i"s accepted in the range of 0–1, must be greater than 0.5, which is the minimum value, in order to support the bidding decision.

Afterward, the bid price for the tender for which decide to bid is determined. We assume that this price is a linear function of tender attractiveness (Fig. 6.3). In this case, C_{max} (maximum tender price) is calculated from the maximum cost prices and the highest profit rate. Assuming a linear function, the objective function C_{min} (minimum tender price) is calculated, where q_i is the quantity of quantities and ci is the unit prices of these quantities (Eq. 6.2).

Minimum bid price

$$\min C : C = \sum_{j=1}^{m} c_j q_j \tag{6.2}$$

Using the formula below (Eq. 6.3), bidding below the actual cost is prevented. In order to make this calculation, the payment program and periodic progress payments and expenses according to production must be made

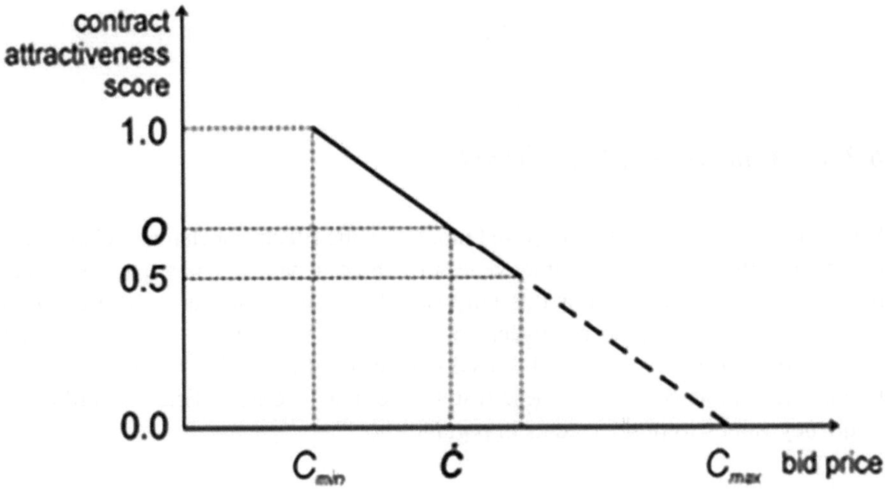

Fig. 6.3 Bid price vs. contract attractiveness

- NPV
- 1/(1 + discount rate) which is discounting factor
- L payment period till the completion date
- Q production quantity
- S, receiving payment from Client per L
- K unit cost
- P, cost payment by Contractor per L

NPV

$$\mathrm{NPV} = \sum_{j=1}^{m}\sum_{l=1}^{t} d^l q_j \left(s_{jl} c_j - p_{jl} k_j\right) \geq 0 \tag{6.3}$$

In addition, bidding unit prices should be selected at a value between minimum and maximum so that there will be no problems in a possible bad situation. For this purpose, it is expected to meet the following condition (Eq. 6.4):

Unit price constraint

$$c_j^{\min} \leq c_j \leq c_j^{\max}, \quad \forall j \in Q \tag{6.4}$$

Finally, the total bid price is distributed among the work quantity breakdowns. For this purpose, the following objective function and constraints are used as another linear programming model (Eq. 6.5 (Biruk et al. (2017)):

Distribution modelling of total bidding price

$$\begin{aligned} &\max \mathrm{NPV} : \mathrm{NPV} = \sum_{j=1}^{m}\sum_{l=1}^{t} d^l q_j \left(s_{jl} \dot{c}_j - p_{jl} k_j\right) \\ &\dot{C} = \sum_{j=1}^{m} \dot{c}_j q_j \\ &c_j^{\min} \leq \dot{c}_j \leq c_j^{\max}, \quad \forall j \in Q \end{aligned} \tag{6.5}$$

6.3.3 *Application of the Model*

Within the framework of the proposed modeling, the Electromechanical Contractor company will evaluate the attractiveness of three separate tender packages that received a request to bid. For this purpose, 10 internal managers and 5 external experts were asked to choose between 45 criteria selected from the literature, giving 1 point to each of 15 criteria. The results were listed according to their selection frequencies (Table 6.2). From these results, the first 15 criteria listed according to frequency values were selected as important.

Table 6.2 Criteria evaluation and 15 selected criteria

İ	Criteria description (factors affecting the bid/no bid decision)	Frequency
1	**Financial capability of the client**	10
2	**Portion can be subcontracted to others**	10
3	**Current workload**	10
4	**General overhead**	10
5	**Requirement of bond capacity**	10
6	**Risk involved in investment**	10
7	**Duration of project**	10
8	**Availability of labour/equipment**	10
9	**Technological difficulty of project beyond the capability of the firm**	10
10	**Need for work**	10
11	**Risk of fluctuation in material price**	10
12	**Availability of capital**	10
13	**Profitability (profit potential)**	9
14	**Value of liquidated damages**	9
15	**Anticipated rate of return**	9
16	Availability of other projects	7
17	Imported materials and equipment	5
18	Degree of hazard/safety	4
19	Tendering duration	4
20	Availability of materials	4
21	Relations with and reputation to client	4
22	Site accessibility	4
23	Method of construction	4
24	Project location	4
25	Site clearance of obstruction	4
26	Experience in similar project	4
27	Prequalification requirement	4
28	Completeness of documents	4
29	Uncertainty due to weather conditions	4
30	Type of owner/client identity	4
31	Project's possible contribution to breaking into new markets	3
32	Tax liability	3
33	Insurance premium	1
34	Fulfilling the "to tender" condition	1
35	Type of contract	1
36	Owner's requirement	2
37	Tendering method	1
38	Bidding document price	1
39	Site condition	1
40	Competitiveness of competitors	0
41	Number of competitors	0

(continued)

Table 6.2 (continued)

İ	Criteria description (factors affecting the bid/no bid decision)	Frequency
42	Chances of obtaining the job	0
43	Project size	0
44	Project type	0
45	Competitive environment	0

6.3.3.1 Defining the Weights of Criteria with AHP and Consistency

The comparative decision matrix (pairwise comparison matrix) of these 15 criteria was then evaluated by 10 decision-makers within the company and the general pairwise comparative decision matrix was created accordingly (Table 6.3).

While creating the comparative decision matrix, the relative importance of the criteria was selected by the evaluators according to the fundamental scale (Table 6.4). Thus, the weight of each criterion was calculated. Consistency testing was carried out by creating a normalized version of these decision matrices (Table 6.5).

The compatibility test for each criterion and the consistency test of the overall matrix were calculated (Table 6.6) according to the Saaty random index (Fig. 6.4).

In AHP, the consistency ratio (CR) is expected to be greater than 0.1 in terms of the reliability of the results. Each evaluator evaluates 15×15 pairwise comparison matrices, and in cases where the consistency ratio (CR) is less than 0.5, the evaluations are reviewed to make them consistent. (In this study, the pairwise comparison matrices made by 10 decision-making evaluators are not presented.)

Finally, total priority weightings are made for the overall problem and the consistency rate is determined. (Table 6.6)

The total consistency ratio CR is 0.06 and is consistent as follows:

	λ_{max}	16.43
$(\lambda_{max} - n)/(n - 1)$	CI	0.10
Random Index	RI	1.58
CI/RI	CR	0.06

6.3.3.2 Tender Attractiveness Score Evaluation and Bid or No Bid

In order to calculate the tender attractiveness scores, 15 evaluators were asked to determine the attractiveness values of the tenders according to 15 selected criteria. These values will range from 1 to 9, ranging from least preferred (1) to most preferred (9). It will also be taken into consideration whether each criterion is maximum or minimum (Table 6.7).

At this stage, the attractiveness scores of the criteria in the projects are weighted using the simple total weighting method (Table 6.8). To do this, first the decision matrix is normalized and the total attractiveness score (Eq. 6.1) is calculated for

Table 6.3 Pairwise comparison matrix

		1	2	3	4	5	6	7	8	9	10	11	12	13	14	15		
	Selected criteria	Financial capability of the client	Portion subcontracted to others	Current workload	General overhead	Requirement of bond capacity	Risk involved in investment	Duration of project	Availability of labor/ equipment	Technological difficulty of project beyond the capability of the firm	Need for work	Risk of fluctuation in material price	Availability of capital	Profitability (profit potential)	Value of liquidated damages	Anticipated rate of return	Total	Priority vector w (average)
1	Financial capability of the client	**1**	3	3	3	1	3	3	5	5	5	5	7	7	7	9	67.00	4.47
2	Portion can be subcontracted to others	0.33	**1**	1	3	3	3	5	5	5	5	5	5	5	7	7	60.33	4.02
3	Current workload	0.33	1	**1**	3	3	3	3	3	5	5	5	7	7	7	9	62.33	4.16
4	General overhead	0.33	0.33	0.33	**1**	1	3	3	3	3	5	5	5	5	7	9	51.00	3.40
5	Requirement of bond capacity	1	0.33	0.33	1	**1**	1	3	3	3	3	5	5	7	7	7	47.67	3.18
6	Risk involved in investment	0.33	0.33	0.33	0.33	1	**1**	1	3	3	5	5	5	7	7	9	48.33	3.22
7	Duration of project	0.33	0.20	0.33	0.33	0.33	1	**1**	1	1	3	3	3	5	7	9	35.53	2.37
8	Availability of labor/ equipment	0.20	0.20	0.33	0.33	0.33	0.33	1	**1**	1	1	3	3	3	3	3	20.73	1.38
9	Tech. difficulty of project …	0.20	0.20	0.20	0.33	0.33	0.33	1	1	**1**	1	3	3	3	3	3	20.60	1.37
10	Need for work	0.20	0.20	0.20	0.20	0.33	0.20	0.33	1	1	**1**	3	3	3	5	7	25.67	1.71

(continued)

Table 6.3 (continued)

		1	2	3	4	5	6	7	8	9	10	11	12	13	14	15		
	Selected criteria	Financial capability of the client	Portion subcontracted to others	Current workload	General overhead	Requirement of bond capacity	Risk involved in investment	Duration of project	Availability of labor/ equipment	Technological difficulty of project beyond the capability of the firm	Need for work	Risk of fluctuation in material price	Availability of capital	Profitability (profit potential)	Value of liquidated damages	Anticipated rate of return	Total	Priority vector w (average)
11	Risk of fluctuation in material price	0.20	0.20	0.20	0.20	0.20	0.20	0.33	0.33	0.33	0.33	**1**	1	1	3	3	11.53	0.77
12	Availability of capital	0.14	0.20	0.14	0.20	0.20	0.20	0.33	0.33	0.33	0.33	1	**1**	1	3	5	13.42	0.89
13	Profitability (profit potential)	0.14	0.20	0.14	0.20	0.14	0.14	0.20	0.33	0.33	0.33	1.00	1	**1**	3	5	13.17	0.88
14	Value of liquidated damages	0.14	0.14	0.14	0.14	0.14	0.14	0.14	0.33	0.33	0.20	0.33	0.33	0.33	**1**	3	6.87	0.46
15	Anticipated rate of return	0.11	0.14	0.11	0.11	0.14	0.11	0.11	0.33	0.33	0.14	0.33	0.20	0.20	0.33	**1.0**	3.72	0.25
	Total	5.0	7.7	7.8	13.4	12.2	16.7	22.5	27.7	29.7	35.3	45.7	49.5	55.5	70.3	89.0		

Table 6.4 Importance scales

Fundamental scale	
0.11	Extremely less important
0.13	
0.14	Very strongly less important
0.17	
0.20	Strongly less important
0.25	
0.33	Moderately less important
0.50	
1	Equal importance
2	
3	Moderately more important
4	
5	Strongly more important
6	
7	Very strongly more important
8	
9	Extremely more important

each alternative tender using the following formulas (Eq. 6.6) according to the desired maximum and minimum states of the criteria.

Normalizing criteria as per desired maximizing or minimazing

$$r_{ij} = \frac{x_{ij}}{x_j^{\max,}}, \quad i = 1,2,\ldots m; \quad j = 1,2,\ldots n \tag{6.6}$$

$$r_{ij} = \frac{x_j^{\min}}{x_{ij}}$$

As can be seen, the calculated total attractiveness scores are greater than 0.5 for each alternative (Table 6.9). These three alternatives are biddable according to the model and their attractiveness ranking according to the criteria is Tender-C > Tender-B ≥ Tender-A and their qualities are as follows:

Tender-A: Industrial Facility 65.000 m^2
Tender-B: Airport Project 210.000 m^2
Tender-C: Hospital Project 150.000 m^2

Within the framework of the calculated attractiveness scores and according to the workload situation, the contractor decided to make an offer for Tender-C.

Table 6.5 Normalized pairwise comparison matrix

		1	2	3	4	5	6	7	8	9	10	11	12	13	14	15		
	Selected criteria	Financial capability of the client	Portion subcontracted to others	Current workload	General overhead	Requirement of bond capacity	Risk involved in investment	Duration of project	Availability of labor/ equipment	Technological difficulty of project beyond the capability of the firm	Need for work	Risk of fluctuation in material price	Availability of capital	Profitability (profit potential)	Value of liquidated damages	Anticipated rate of return	Total	Average (final priorities) w
1	Financial capability of the client	**0**	0	0	0	0	0	0	0	0	0	0	0	0	0	0	2.66	0.18
2	Portion can be subcontracted to others	0.07	**0**	0	0	0	0	0	0	0	0	0	0	0	0	0	2.17	0.14
3	Current workload	0.07	0	**0**	0	0	0	0	0	0	0	0	0	0	0	0	2.11	0.14
4	General overhead	0.07	0.04	0.04	**0**	0	0	0	0	0	0	0	0	0	0	0	1.48	0.10
5	Requirement of bond capacity	0	0.04	0.04	0	**0**	0	0	0	0	0	0	0	0	0	0	1.45	0.10
6	Risk involved in investment	0.07	0.04	0.04	0.02	0	**0**	0	0	0	0	0	0	0	0	0	1.25	0.08
7	Duration of project	0.07	0.03	0.04	0.02	0.03	0	**0**	0	0	0	0	0	0	0	0	0.86	0.06
8	Availability of labor/ equipment	0.04	0.03	0.04	0.02	0.03	0.02	0	**0**	0	0	0	0	0	0	0	0.58	0.04
9	Tech. difficulty of project …	0.04	0.03	0.03	0.02	0.03	0.02	0	0	**0**	0	0	0	0	0	0	0.56	0.04
10	Need for work	0.04	0.03	0.03	0.01	0.03	0.01	0.01	0	0	**0**	0	0	0	0	0	0.59	0.04

		1	2	3	4	5	6	7	8	9	10	11	12	13	14	15		
	Selected criteria	Financial capability of the client	Portion subcontracted to others	Current workload	General overhead	Requirement of bond capacity	Risk involved in investment	Duration of project	Availability of labor/ equipment	Technological difficulty of project beyond the capability of the firm	Need for work	Risk of fluctuation in material price	Availability of capital	Profitability (profit potential)	Value of liquidated damages	Anticipated rate of return	Total	Average (final priorities) w
11	Risk of fluctuation in material price	0.04	0.03	0.03	0.01	0.02	0.01	0.01	0.01	0.01	0.01	**0**	0	0	0	0	0.32	0.02
12	Availability of capital	0.03	0.03	0.02	0.01	0.02	0.01	0.01	0.01	0.01	0.01	0	**0**	0	0	0	0.32	0.02
13	Profitability (profit potential)	0.03	0.03	0.02	0.01	0.01	0.01	0.01	0.01	0.01	0.01	0.02	0	**0**	0	0	0.31	0.02
14	Value of liquidated damages	0.03	0.02	0.02	0.01	0.01	0.01	0.01	0.01	0.01	0.01	0.01	0.01	0.01	**0**	0	0.20	0.01
15	Anticipated rate of return	0.02	0.02	0.01	0.01	0.01	0.01	0.00	0.01	0.01	0.00	0.01	0.00	0.00	0.00	**0.0**	0.14	0.01
	Total	1.0	1.0	1.0	1.0	1.0	1.0	1.0	1.0	1.0	1.0	1.0	1.0	1.0	1.0	1.0	15.0	1.0

Table 6.6 Consistency test matrix

		1	2	3	4	5	6	7	8	9	10	11	12	13	14	15			
	Selected criteria	Financial capability of the client	Portion subcontracted to others	Current workload	General overhead	Requirement of bond capacity	Risk involved in investment	Duration of project	Availability of labor/ equipment	Technological difficulty of project beyond the capability of the firm	Need for work	Risk of fluctuation in material price	Availability of capital	Profitability (profit potential)	Value of liquidated damages	Anticipated rate of return	Normalized matrix total	Consistency measure	Eigen value (λ)
1	Financial capability of the client	**1**	3	3	3	1	3	3	5	5	5	5	7	7	7	9	2.66	45.08	1.13
2	Portion can be subcontracted to others	0.33	**1**	1	3	3	3	5	5	5	5	5	5	5	7	7	2.17	37.83	1.16
3	Current workload	0.33	1	**1**	3	3	3	3	3	5	5	5	7	7	7	9	2.11	36.49	1.16
4	General overhead	0.33	0.33	0.33	**1**	1	3	3	3	3	5	5	5	5	7	9	1.48	25.41	1.15
5	Requirement of bond capacity	1	0.33	0.33	1	**1**	1	3	3	3	3	5	5	7	7	7	1.45	23.83	1.10
6	Risk involved in investment	0.33	0.33	0.33	0.33	1	**1**	1	3	3	5	5	5	7	7	9	1.25	20.81	1.11
7	Duration of project	0.33	0.20	0.33	0.33	0.33	1	**1**	1	1	3	3	3	5	7	9	0.86	14.19	1.10
8	Availability of labor/ equipment	0.20	0.20	0.33	0.33	0.33	0.33	1	**1**	1	1	3	3	3	3	3	0.58	9.54	1.10
9	Tech. difficulty of project …	0.20	0.20	0.20	0.33	0.33	0.33	1	1	**1**	1	3	3	3	3	3	0.56	9.26	1.10

		1	2	3	4	5	6	7	8	9	10	11	12	13	14	15			
	Selected criteria	Financial capability of the client	Portion subcontracted to others	Current workload	General overhead	Requirement of bond capacity	Risk involved in investment	Duration of project	Availability of labor/ equipment	Technological difficulty of project beyond the capability of the firm	Need for work	Risk of fluctuation in material price	Availability of capital	Profitability (profit potential)	Value of liquidated damages	Anticipated rate of return	Normalized matrix total	Consistency measure	Eigen value (λ)
10	Need for work	0.20	0.20	0.20	0.20	0.33	0.20	0.33	1	1	**1**	3	3	3	5	7	0.59	9.30	1.05
11	Risk of fluctuation in material price	0.20	0.20	0.20	0.20	0.20	0.20	0.33	0.33	0.33	0.33	**1**	1	1	3	3	0.32	5.07	1.06
12	Availability of capital	0.14	0.20	0.14	0.20	0.20	0.20	0.33	0.33	0.33	0.33	1	**1**	1	3	5	0.32	5.09	1.05
13	Profitability (profit potential)	0.14	0.20	0.14	0.20	0.14	0.14	0.20	0.33	0.33	0.33	1.00	1	**1**	3	5	0.31	4.82	1.04
14	Value of liquidated damages	0.14	0.14	0.14	0.14	0.14	0.14	0.14	0.33	0.33	0.20	0.33	0.33	0.33	**1**	3	0.20	3.16	1.06
15	Anticipated rate of return	0.11	0.14	0.11	0.11	0.14	0.11	0.11	0.33	0.33	0.14	0.33	0.20	0.20	0.33	**1.0**	0.14	2.35	1.08
	Total	5.0	7.7	7.8	13.4	12.2	16.7	22.5	27.7	29.7	35.3	45.7	49.5	55.5	70.3	89.0		252.2	16.43

Random Index (RI)

n	1	2	3	4	5	6	7	8	9	10	11	12	13	14	15
RI	0.00	0.00	0.58	0.90	1.12	1.24	1.32	1.41	1.45	1.49	1.51	1.48	1.56	1.57	1.58

Fig. 6.4 Saaty random indices

6.3.3.3 Linear Modeling of Calculating Total Bid Price

The maximum bid price (C_{max}) will be calculated based on the highest resource prices in the market and the highest overhead and profit. The minimum tender price (C_{min}), which is the objective function in modeling, will be calculated with the linear programming modeling given in the method.

According to the company database and current prices, the actual cost was calculated by the bidding department as 405,000,000 SAR and the maximum industry cost as 465,512,000 SAR. In addition, a work plan and progress payment schedule were prepared in accordance with the project duration (Tables 6.10 and 6.11).

Discount factors in accordance with employer payment shares and contractor cost payment shares were determined in each period when the opportunity rate was selected as 1% (Table 6.12).

With this information, the objective function C_{min} (Eq. 6.2), decision variables bid unit prices c_j and constraints are solved in Excel Solver, provided that NPV (Eq. 6.3) and c_j remain between the market minimum and maximum unit prices (Eq. 6.4) according to the progress payment and payment plans. The minimum bid price was SAR 418,960,800 and the NPV was SAR 13,822,575.24.

6.3.3.4 Distributing Total Bid Price to WBS

Assuming that the bid price is a linear function of the attractiveness score and by interpolation (Eq. 6.5), Cdot was calculated as 427,805,528 SAR and distributed to the quantities (Table 6.13).

6.4 Conclusion

Decisions about bidding tenders and at what price directly affect the contractor's vision of being a sustainable and profitable company. Therefore, it is not very rational to act only intuitively and based on experience when making these decisions. To avoid losses, the contractor must act with reliable inputs and reliable solutions in the relevant decision problem.

In this study, it is aimed to study AHP and Linear programming methods and theorems to select the most beneficial invitation among the tenders, and to calculate

Table 6.7 Attractiveness scores of alternatives

Tender bidding criteria and scores																						
No	Selected criteria	A	B	C	Expect	Value	1	2	3	4	5	6	7	8	9	10	11	12	13	14	15	Total	*W*
1	Financial capability of the client	6	9	8	max	9	9	9	9	9	9	9	9	9	9	9	9	9	9	9	9	135	1
2	Portion can be subcontracted to others	7	3	1	max	7	7	8	6	7	6	8	8	9	9	8	7	7	6	7	8	111	1
3	Current workload	5	5	5	min	5	9	8	8	9	8	8	9	8	7	8	8	6	7	6	8	117	1
4	General overhead	5	3	1	min	1	7	6	6	8	8	5	7	6	5	6	7	5	5	6	7	94	1
5	Requirement of bond capacity	6	4	2	min	2	6	6	7	8	8	7	6	5	5	4	5	6	5	6	5	89	1
6	Risk involved in investment	7	5	3	min	3	6	4	5	7	5	6	4	4	7	5	8	4	3	6	4	78	1
7	Duration of project	5	8	6	min	5	7	4	6	5	6	7	4	5	6	6	5	5	7	3	3	79	1
8	Availability of labor/equipment	8	5	2	max	8	8	9	9	5	2	3	5	7	6	8	7	5	6	7	8	95	1
9	Technological difficulty of project	8	4	2	min	2	8	7	9	6	5	6	6	7	7	5	4	4	5	5	4	88	1
10	Need for work	9	4	1	min	1	8	9	8	7	8	7	8	7	6	7	5	6	6	5	4	101	1
11	Risk of fluctuation in material price	7	6	2	min	2	8	7	6	6	6	7	7	7	6	4	5	5	6	9	7	96	1
12	Availability of capital	4	4	4	max	4	8	7	7	7	7	6	7	5	5	5	5	7	6	7	8	97	1
13	Profitability (profit potential)	5	8	7	max	8	9	8	8	7	8	7	9	9	9	7	6	6	6	8	7	114	1
14	Value of liquidated damages	7	5	3	min	3	6	6	6	6	7	5	8	6	4	9	6	5	7	6	8	95	1
15	Anticipated rate of return	5	9	8	max	9	9	9	8	7	8	7	5	4	5	6	6	5	6	7	7	99	1
							115	107	108	104	101	98	102	98	96	97	93	85	90	97	97	1488	15

Table 6.8 Criteria attractiveness score of alternatives

Tender bidding criteria and scores				
No	Selected criteria	A	B	C
1	Financial capability of the client	0.67	1.00	0.89
2	Portion can be subcontracted to others	1.00	0.43	0.14
3	Current workload	1.00	1.00	1.00
4	General overhead	0.20	0.33	1.00
5	Requirement of bond capacity	0.33	0.50	1.00
6	Risk involved in investment	0.43	0.60	1.00
7	Duration of project	1.00	0.63	0.83
8	Availability of labor/equipment	1.00	0.63	0.25
9	Technological difficulty of project beyond the capability of the firm	0.25	0.50	1.00
10	Need for work	0.11	0.25	1.00
11	Risk of fluctuation in material price	0.29	0.33	1.00
12	Availability of capital	1.00	1.00	1.00
13	Profitability (profit potential)	0.63	1.00	0.88
14	Value of liquidated damages	0.43	0.60	1.00
15	Anticipated rate of return	0.56	1.00	0.89

Table 6.9 Total attractiveness score of alternatives

Tender bidding criteria and scores				
No	Selected criteria	A	B	C
1	Financial capability of the client	0.12	0.18	0.16
2	Portion can be subcontracted to others	0.14	0.06	0.02
3	Current workload	0.14	0.14	0.14
4	General overhead	0.02	0.03	0.10
5	Requirement of bond capacity	0.03	0.05	0.10
6	Risk involved in investment	0.04	0.05	0.08
7	Duration of project	0.06	0.04	0.05
8	Availability of labor/equipment	0.04	0.02	0.01
9	Technological difficulty of project beyond the capability of the firm	0.01	0.02	0.04
10	Need for work	0.00	0.01	0.04
11	Risk of fluctuation in material price	0.01	0.01	0.02
12	Availability of capital	0.02	0.02	0.02
13	Profitability (profit potential)	0.01	0.02	0.02
14	Value of liquidated damages	0.01	0.01	0.01
15	Anticipated rate of return	0.01	0.01	0.01
	Total	0.65	0.67	0.81

Table 6.10 Provisional payment percentages from client

	sjl (client share)														
WBS	1	2	3	4	5	6	7	8	9	10	11	12	13	14	15
Fire fighting			5%	5%	10%	20%	20%	20%	10%	5%					5%
Hanger and supports	5%	10%	10%	15%	15%	10%	10%	10%	5%	5%					5%
Vibration control						5%	10%	15%	15%	15%	15%	10%	10%		5%
Identification							5%	10%	15%	15%	15%	15%	10%	10%	5%
Drainage system	5%	5%	5%	10%	15%	10%	10%	10%	5%	5%	5%	5%	5%		5%
Plumbing system	5%	5%	5%	10%	15%	10%	10%	10%	5%	5%	5%	5%	5%		5%
Heating cooling system		5%	5%	5%	10%	15%	10%	10%	10%	5%	5%	5%	5%	5%	5%
HVAC system		5%	5%	5%	10%	15%	10%	10%	10%	5%	5%	5%	5%	5%	5%
Water distribution system	5%	5%	5%	10%	15%	10%	10%	10%	5%	5%	5%	5%	5%		5%
Hydronic system	5%	10%	10%	15%	20%	10%	10%	10%	5%						5%
Steam system								10%	10%	15%	25%	25%	10%		5%
Insulation works	5%	5%	5%	10%	15%	10%	10%	10%	5%	5%	5%	5%	5%		5%
Valves					5%	5%	15%	15%	15%	15%	10%	10%	5%		5%
Meters and gauges						5%	5%	15%	15%	15%	15%	10%	10%	5%	5%
Plumbing fixtures									5%	10%	15%	25%	25%	15%	5%
Building management system									5%	5%	15%	25%	25%	20%	5%
Fire alarm system			5%	5%	10%	20%	20%	20%	10%	5%					5%
Lightning system										10%	20%	25%	25%	15%	5%
Grounding system			5%	5%	5%	5%	5%	10%	10%	10%	10%	10%	10%	10%	5%
Cable trunking system	5%	5%	5%	10%	10%	10%	10%	10%	10%	10%	5%	5%			5%
LV cabling					10%	20%	20%	20%	20%	5%					5%
MV cabling					10%	25%	25%	15%	15%	5%					5%
Lighting system					5%	5%	5%	10%	10%	15%	15%	15%	10%	5%	5%
Wiring devices					5%	5%	5%	10%	10%	15%	15%	15%	10%	5%	5%
LV distribution system									15%	20%	20%	20%	10%	10%	5%
MV distribution system									10%	20%	25%	25%	10%	5%	5%
ELV systems								5%	15%	15%	15%	15%	15%	15%	5%
ICT system									5%	10%	20%	25%	25%	10%	5%

Table 6.11 Provisional contractor's cost percentages

	pjl (contractor share)														
WBS	1	2	3	4	5	6	7	8	9	10	11	12	13	14	15
Fire fighting	10%		40%	40%	10%										
Hanger and supports	10%	15%	15%	15%	15%	15%	15%								
Vibration control			10%				90%								
Identification	10%				25%	25%	25%	25%							
Drainage system	5%	10%		25%	25%	25%	10%								
Plumbing system	10%	10%		20%	20%	20%	20%								
Heating cooling system	5%	10%		25%	25%	25%	10%								
HVAC system	10%	10%	20%	20%	20%	10%	10%								
Water distribution system	10%	10%	10%	20%	20%	20%	10%								
Hydronic system	10%	10%	40%	40%											
Steam system	10%							60%	30%						
Insulation works	10%	20%	20%	30%	20%										
Valves		10%	10%	40%	40%										
Meters and gauges		50%			50%										
Plumbing fixtures			20%						30%	30%	20%				
Building management system						0.2			20%	40%	10%	10%			
Fire alarm system	10%	20%			40%	20%	10%								
Lightning system				10%						90%					
Grounding system	10%			90%											
Cable trunking system	10%	20%	20%	20%	20%	10%									
LV cabling	50%				30%	10%	10%								

	pjl (contractor share)														
WBS	1	2	3	4	5	6	7	8	9	10	11	12	13	14	15
MV cabling	50%				40%	10%									
Lighting system			10%	20%		20%	30%	10%	10%						
Wiring devices			20%			40%			40%						
LV distribution system						50%			20%	10%	10%	10%			
MV distribution system			50%			40%	10%								
ELV systems		10%	10%	10%				20%	20%	20%	10%				
ICT system						0.25			20%	25%	15%	15%			

Table 6.12 Discounting factor and discount rate

Opportunity cost (r)	1%	1%	1%	1%	1%	1%	1%	1%	1%	1%	1%	1%	1%	1%	1%
dl (discount rate)	0.99	0.99	0.99	0.99	0.99	0.99	0.99	0.99	0.99	0.99	0.99	0.99	0.99	0.99	0.99

Table 6.13 Distributing bid price

j	WBS	Unit	Quantity q	Unit cost kj	Min. unit price c_{jmin}	Max. unit price c_{jmax}	Bid unit price c_{jdot}
1	Fire fighting	m^2	150,000.00	126.23	130.58	145.08	133.33
2	Hanger and supports	m^2	150,000.00	81.68	84.49	93.88	86.27
3	Vibration control	m^2	150,000.00	5.94	6.14	6.83	6.27
4	Identification	m^2	150,000.00	7.43	7.68	8.53	7.84
5	Drainage system	m^2	150,000.00	59.40	61.45	68.28	62.74
6	Plumbing system	m^2	150,000.00	103.95	107.53	119.48	109.80
7	Heating cooling system	m^2	150,000.00	163.35	168.98	187.76	172.55
8	HVAC system	m^2	150,000.00	207.90	215.07	238.96	219.61
9	Water distribution system	m^2	150,000.00	118.80	122.90	136.55	125.49
10	Hydronic system	m^2	150,000.00	193.05	199.70	221.89	203.92
11	Steam system	m^2	150,000.00	111.38	115.21	128.02	117.65
12	Insulation works	m^2	150,000.00	29.70	30.72	34.14	31.37
13	Valves	m^2	150,000.00	22.28	23.04	25.60	23.53
14	Meters and gauges	m^2	150,000.00	8.91	9.22	10.24	9.41
15	Plumbing fixtures	m^2	150,000.00	178.20	184.34	204.83	188.23
16	Building management system	m^2	150,000.00	66.83	69.13	76.81	70.59
17	Fire alarm system	m^2	150,000.00	97.20	100.55	111.72	102.67
18	Lightning system	m^2	150,000.00	3.65	3.77	4.19	3.85
19	Grounding system	m^2	150,000.00	6.08	6.28	6.98	6.42
20	Cable trunking system	m^2	150,000.00	151.88	157.11	174.57	160.43

(continued)

Table 6.13 (continued)

j	WBS	Unit	Quantity q	Unit cost kj	Min. unit price c_{jmin}	Max. unit price c_{jmax}	Bid unit price c_{jdot}
21	LV cabling	m^2	150,000.00	182.25	188.53	209.48	192.51
22	MV cabling	m^2	150,000.00	66.83	69.13	76.81	70.59
23	Lighting system	m^2	150,000.00	115.43	119.40	132.67	121.92
24	Wiring devices	m^2	150,000.00	18.23	18.85	20.95	19.25
25	LV distribution system	m^2	150,000.00	130.01	134.49	149.43	137.33
26	MV distribution system	m^2	150,000.00	66.83	69.13	76.81	70.59
27	ELV systems	m^2	150,000.00	243.00	251.38	279.31	256.68
28	ICT system	m^2	150,000.00	133.65	138.26	153.62	141.18
	Total	m^2	150,000	405,000,000	418,960,800	465,512,000	427,805,528

the bid price of the selected tender and distribute it to the quantities. The study was carried out in three stages. In the first stage, the selection of the 15 most important criteria among 45 criteria was determined by the ratings of 15 evaluators, and these selected criteria were weighted with the data of 10 evaluators using the AHP method. At the same time, the decision to bid/no-bid was made by determining the benefit attractiveness scores of these 15 most important selected criteria with the data of 15 evaluators. In the second stage, the minimum bid price and final bid price were created with the help of the linear programming method and Excel Solver used for the bid decision. In the final stage, this bid price was distributed to the quantities.

Within the framework of the modeling, we propose in the study, the Electromechanical Contractor company will be able to create the bid price with an analytical approach by making a rational-based choice among the tenders for which proposals are requested. This modeling, which is quite sufficient in terms of the bid/no bid decision, can prevent the company from incurring losses in terms of creating the bid price and allow it to enter the tender by exceeding a reasonable time limit. However, in cases where there is sufficient time, it is of course better for the company to conduct detailed cost analysis to determine the tender offer price. For this purpose, the establishment and correct operation of internal decision processes and decision support systems are inevitable for profitable business and sustainable growth.

Acknowledgment This study was undertaken as part of the MBA program in the Graduate School of Business in the School of Business at Istanbul University, within the framework of a graduation project conducted by Cihat Doğan, MBA, and supervised by Prof. Dr. Eyüp Çetin. The data, expert evaluations and cost calculations used in the study have been utilized with the knowledge and permission of Sasel Saudi L.L.C. and we extend our thanks to them. We would like to thank Series Editor Prof. Dr. Kıymet Çalıyurt for her interest and support.

References

Ahmad I, Minkarah I (1988) Questionnaire survey on bidding in construction. J Manag Eng 4(3):229–243
Bageis S, Fortune C (2009) Factors affecting the bid/no bid decision in the Saudi Arabian construction contractors. Constr Manag Econ 27(1):53–71
Biruk S, Jaskowski P, Czarnigowska A (2017) Modeling contractor's bidding decisions. Procedia Eng 182:91–98
Boran FE et al (2009) A multi-criteria intuitionistic fuzzy group decision making for supplier selection with TOPSIS method. Expert Syst Appl 36:11363–11368
Çetin E, Kiremitci B (2018) Pride and passion: remarks on Thomas L. Saaty's final papers. Eur J Pure Appl Math 11(3):893–897
Cheng M-Y et al (2011) Bidding decision making for construction company using multi-criteria prospect model. J Civil Eng Manag 17(3):426–434
Chua DKH, Li D (2000) Key factors in bid reasoning model. J Constr Eng Manag 126:349–357
Cui DH, Zhang XY (2009) Application of game theory on bidding price decision. In: IEEE 16th International Conference on Industrial. IEEE, Beijing, pp 58–61
Jato-Espino D et al (2014) A review of application of multi-criteria decision making methods in construction. Autom Const 45:151–162
Köksalan MM, Wallenius J, Zionts S (2011) Multiple criteria decision making: from early history to the 21st century. World Scientific Publishing Company, Singapore
Leśniak A, Plebankiewicz E (2015) Modeling the decision-making process concerning participation in construction bidding. J Manag Eng 31(2)
Li H, Shen LY, Love PE (1999) ANN-based mark-up estimation system with self-explanatory capacities. J Constr Eng Manag 125(3):185–189
Mielec A et al (2009) Determining profit level: the contractor's bidding strategy, vol 6. Przegląd Budowlany, pp 50–53
Oo B-L et al (2007) Applying a random coefficients logistic model to contractors ' decision to bid. Constr Manag Econ 25:387–398
Oyeyipo OO, Odusami KT, Ojelabi RA, Afolabi AO (2016) Factors affecting contractors' bidding decision for construction projects in Nigeria. J Constr Dev Countr 21(2):21–35
Saaty T (1994) Fundamentals of decision making and priority theory with the analytic hierarchy process. RWS Publications, Pittsburgh, PA
Saaty T (2008) The analytic hierarchy and analytic network measurement processes: applications to decisions under risk. Eur J Pure Appl Math 1(1):122–196
Shash AA (2000) Factors considered in tendering decisions by top UK contractors. Constr Manag Econ 11:111–118
Triantaphyllou E (2000) Multi-criteria decision making methods: a comparative study. Springer, Boston Kluwer
Vahdani B et al (2010) Extension of VIKOR method based on interval-valued fuzzy sets. Adv Manuf Technol 47:1231–1239
Wang WC et al (2007) Integration of simulation-based cost model and multi-criteria evaluation model for bid price decisions. Comput Aided Civ Infrastruct Eng 22:223–235
Wanous M, Boussabaine AH, Lewis J (2000) To bid or not to bid: a parametric solution. Constr Manag Econ 18(4):457–466

Cihat Doğan was born in 1975 in Kayseri and received his high school education in Ankara Bahçelievler Cumhuriyet Lisesi. He under-graduated Electrical & Electronics Engineering at METU and Business Administration at Istanbul University, respectively. Later, he completed an MBA at Istanbul University. Currently continuing his education as a senior student in the Department of Sociology at Istanbul University. He has been working in the field of international construction contracting for 24 years and continues to serve as a senior director in a private company.

Eyüp Çetin is a tenured Professor of Operations Research and Statistics in the Department of Quantitative Methods at the School of Business, Istanbul University, and an Affiliate Distinguished Professor at York University, Canada. He is also affiliated with New York Business Global, USA. He has held various managerial positions. He is former and founding Dean of the Faculty of Transportation and Logistics and former Vice Dean of the School of Business at Istanbul University. He received a Ph.D. in quantitative methods from Istanbul University in 2004. He has taught courses in management science, statistics, business analytics, stochastic models, game theory and health analytics to all levels of business, engineering, healthcare management students: undergraduates, MBAs and Ph.D. students. He has been a consultant for numerous companies, and he has taught statistical & decision modelling and business analytics in even global companies such as Vodafone. His current research focuses on business analytics, data science, digital business & health, healthcare management science, mathematical medicine, operations research, mathematical & statistical modelling. Dr. Çetin is also Founder and Editor-in-Chief of prestigious international academic journals such as European Journal of Pure & Applied Mathematics.

Chapter 7
Entrepreneurship Talent or Skill?

Vesna Novaković, Elvis Mujkić, and Dragan Milovanović

Abstract At the end of the twentieth century and the beginning of the twenty-first century, entrepreneurship emerged as one of the leading drivers of the economic development of a country. This research clarifies whether entrepreneurship can be learned or if potential entrepreneurs are born with an innate talent for entrepreneurship. Our research addresses the problem we defined with the question: Are entrepreneurial abilities and skills innate in a person or are they acquired over time? Based on this, we formulated the hypothesis: Entrepreneurship is a set of skills and knowledge that can be acquired, which we will confirm or refute through the presentation of the theoretical introduction and research results. The subject of research in this chapter is the attitudes of entrepreneurs regarding whether entrepreneurship can be learned by mastering entrepreneurial skills or whether entrepreneurs believe it is a talent.

Keywords Economics of education · Teaching of economics · Graduate · Education and economic development · Entrepreneurship · Management

---

V. Novaković (✉)
Faculty of Business Economy, Paneuroperan University Apeiron, Banja Luka, Bosnia and Herzegovina

E. Mujkić
Banja Luka, Bosnia and Herzegovina

D. Milovanović
Faculty of Economics, University of Banja Luka, Banja Luka, Bosnia and Herzegovina
e-mail: dragan.milovanovic@ef.unibl.org

K. T. Çalıyurt (ed.), *New Approaches to CSR, Sustainability and Accountability, Volume VI*, Accounting, Finance, Sustainability, Governance & Fraud: Theory and Application, https://doi.org/10.1007/978-981-96-5937-1_7

7.1 Introduction

The subject of modern economic thought dealing with microeconomics is largely occupied with entrepreneurship: the impact of entrepreneurship on the development and stimulation of the economy in less developed countries and its effect on reducing the unemployment rate. What is entrepreneurship? To gain a clear understanding of the role of entrepreneurship, it must be considered in a broader social, cultural, and economic context. Entrepreneurship can be seen as a synonym with introducing something new, different, specific, or unique. Entrepreneurship represents willingness for action, it represents initiative, perseverance, persistence, energy, willingness to take risk, and adapt to change (Novaković 2018).

The question we posed in this research, namely, the idea of learning and teaching entrepreneurship, is a constant debate in academic and entrepreneurial circles. When examining the status of entrepreneurs among young people, it is emotionally attractive and reflects a successful individual.

Positive examples from entrepreneurship encourage young people to start their own business ventures. However, the question arises: Can entrepreneurship be learned?

The World Economic Forum's "The Future of Jobs Report" annual report shows a significant gap between the skills needed by employers and the skills young people acquire after completing their education. Acquiring new skills, competences, and knowledge is the only guarantee for obtaining quality jobs where young people can fulfill their professional desires and goals and contribute to innovation and economic growth in their country. Analytical thinking and creative thinking remain the most important skills for workers in 2023. Analytical thinking is considered a fundamental skill by more companies than any other skill. Creative thinking, another cognitive skill, is in second place. The next three self-efficacy skills are: resilience, flexibility, and agility. Reliability and attention to detail are ranked sixth, behind technological literacy (World Economic Forum 2023).

The development of soft skills, such as problem solving, critical thinking, emotional intelligence development, curiosity and creativity, is in high demand (and necessary) because these skills are difficult to replace with artificial intelligence or robotics (Nagy 2020).

According to the opinions of people from practice, universities do not adequately prepare candidates for a career in entrepreneurship, in the sense that education can even prevent young people from becoming successful entrepreneurs (authors research).

If the educational process includes learning about behavior in order to avoid risk, the career of young people will be directed toward employment in secure companies, preferably in state-owned companies. One of the incentives for such behavior may also be a study that has shown that for the segment of workers with secondary vocational education, average wages in the public sector are significantly higher than the average wages in the private sector (34.9%) in the Republic of Srpska (Novakovic and Sašić 2015).

This study will show the attitudes of entrepreneurs based on their experience, regarding whether entrepreneurship can be learned or if an entrepreneur is born, as well as the recommendation of entrepreneurs on which direction young people should develop their competencies and which qualities they should possess.

We base our study on entrepreneurship education conducted in Bosnia and Herzegovina, as well as the opinions of entrepreneurs on whether entrepreneurship is a talent or can be learned. Our research question is as follows: Are entrepreneurial abilities and skills innate in a person, or are they acquired over time? The paper is structured as follows: in the introduction, we first acknowledge the need for studying entrepreneurship and the development of entrepreneurial competencies. In the second part, we discuss the literature on entrepreneurship education. The method we used for research was surveying entrepreneurs through a questionnaire that reflects their views on this topic. In the results section, we show the results of our research, while in the discussion and future research section, we outline what, in our opinion, should be done to prepare young people for the free market better.

7.2 Concept of Entrepreneurship

It is believed that Richard Cantillon was the first scholar who defined entrepreneurship in his work "Essay on the Nature of Trade in General" (Baletić 1982). He describes an entrepreneur as someone willing to take the financial risk of a business venture to create a profit. This definition reflects the role of entrepreneurship in Europe during the eighteenth century, when his work was written. The indispensable author who established entrepreneurship as a driver of economic changes is Jozef Šumpeter (Brown and Thornton 2013). In the twentieth century, he connected the dynamic aspect of entrepreneurship with innovation and economic development. According to Schumpeter, economic growth occurs in the following five cases:

1. Introducing a new product or improving an existing one;
2. Introducing a new production method or improving the existing one;
3. Creating a new sales market, especially the export market;
4. Introducing new raw materials, materials, and semi-finished products; and
5. Creating a new organization or improving the existing one.

Competition constantly imposes the need to improve both technology and organization in companies. As Schumpeter once said: "Change is the only constant in the evolution and development of the capitalist economy", while innovation itself sets the company on the "waves of creative destruction". This last phrase was taken by Schumpeter from Werner Sombart (Novaković 2018). Schumpeter portrayed the entrepreneur as someone who breaks the lull and brings economic changes.

Entrepreneurship represents life attitudes, the willingness to act in order to achieve a vision, it shows the willingness to take risks. In order to create new values, it is necessary to have a clear vision of what is to be achieved over a longer period. Entrepreneurship requires dedication, spending time and effort, undertaking risky ventures, and the most important part is getting a certain reward for everything invested. The reward is independence, followed by personal satisfaction and making a profit as a measure of success. Every person is a potential entrepreneur (Hisrich et al. 2010). The review of the literature shows us that there is still no uniform opinion of scientists as to whether entrepreneurship is a talent or entrepreneurship is a skill that can be acquired and mastered during the educational process. This chapter supports the opinion that entrepreneurship can be mastered, and not that an entrepreneur is born. Therefore, it is necessary to have systemic support for entrepreneurship to stimulate creativity and readiness for risky ventures. The entrepreneur must also be operational, i.e., he must be able to carry out the planned activities, both independently and with the coordination and organization of immediate executors, i.e., with the ability to transfer his enthusiasm and vision to his associates, in order to achieve the desired success through their engagement.

Nowadays, an entrepreneur is considered a person who has decided to start a business venture or has been engaged in starting a business (Popescu, Entrepreneurship—Some General Knowledge, 2013). Just as he creates ideas, he implements them and realizes them, encountering many obstacles in order to reach his set goal. An entrepreneur must possess many skills to be the first to recognize a chance, an opportunity, either through insight or intuition.

Research shows that education is extremely important in terms of an entrepreneur's ability to deal with the problems encountered during the business process, as well as adapting to new situations. Research shows that the wider knowledge base possessed by human capital enables a wider network for discovering and generating potential opportunities (Hisrich et al. 2010).

Hisrich gives us an overview of the skills required by entrepreneurship. Skills are grouped into three units:

1. Technical skills that include: verbal communication, monitoring the environment, ability to organize, ability to build a network of acquaintances, management style, etc.
2. Business management skills include: planning and setting goals, ability to make decisions, marketing, finance, accounting, management, control, negotiation skills, growth management, etc.
3. The characteristics of an entrepreneur that distinguish him as a person are: the ability of internal control, willingness to take risks, innovation, oriented toward changes, perseverance, persistence, he is a leader who has the ability to manage changes.

Personal characteristics, interests and individual strength of an individual also determine the motivations for engaging in entrepreneurship. Whether a person will squander his opportunities or use them and encourage and improve them affects its decisions about how to use opportunities, whether to work hard or to be idle, whether to accept one job or another, whether to engage in an enterprise or not, to save or to spend earned money (Fridman and Fridman 1996).

Aulet states (Aulet 2013) that there are three myths about entrepreneurship that need to be changed or refuted. The first myth is that individuals found companies and that the entrepreneur is a lone hero. However, research shows that successful companies are born thanks to teamwork. Another myth is that all entrepreneurs are charismatic and that their charisma is the key to success. Research shows that charisma is difficult to sustain, although it contributes to success, the greater importance for an entrepreneur is the ability to communicate and success in sales. The third myth is that there is an entrepreneurial gene and that some people are simply destined to be entrepreneurs. However, such a gene does not exist and will not be found. There are real skills that increase the odds of success, and these skills can be learned. People can adapt and learn new behaviors, new processes, new procedures. According to that opinion, entrepreneurship can be divided into individual procedures, processes, behavior models, learning models that can be mastered (Aulet 2013). His evidence for the point of views expressed are the students who attend the Massachusetts Institute of Technology. Namely from 2006 until 2013, over 900 new companies are founded almost every year by students who graduated from the Massachusetts Institute of Technology, these companies employ over three million people, and have a total annual income of over two billion dollars, which makes them the eleventh largest economy in the world.

7.3 Entrepreneurship as Part of the Educational System

Entrepreneurship education has been a topic of growing interest in recent years. Various educational programs and institutions have been established worldwide to promote entrepreneurship and develop entrepreneurial skills among young people. One of the most common reasons why entrepreneurship is encouraged in the educational system is that entrepreneurship is one of the main drivers of economic growth, innovation, and job creation (Badulescu and Badulescu 2013). Also, entrepreneurship can play a significant role as a carrier of changes in society, and entrepreneurship education is encouraged in order to create added value in the public sector as well (Lackéus 2015). The "Small Business Act"document, establishing a comprehensive policy framework for supporting small and medium-sized enterprises, places special emphasis on the importance of entrepreneurial learning and entrepreneurial competencies (European Union 2023). This is particularly emphasized in the "New Skills Program"(European Commission 2023). In most European countries, entrepreneurship education is present through all three levels of education, from primary to higher education (Novakovic and Jakupovic 2018). However, the authors emphasize the difference between teaching about entrepreneurship and teaching through entrepreneurship. Although competencies are acquired through the educational system, entrepreneurship is also considered a practice that can be learned and taught (Jussila et al. 2020). The purpose of entrepreneurial education is to train students to acquire skills, ideas and managerial abilities, as well as capacities for self-employment, not to be employed for a salary. Another goal is to help students consider business as a career by developing positive attitudes toward entrepreneurship (Shah 2020). Entrepreneurship education is focused on content, on theory, while education through entrepreneurship improves the student's ability to perform entrepreneurial actions, i.e., to acquire practical skills (Jussila et al. 2020).

Approaches and goals of entrepreneurship education may differ depending on the context of the individual country and the understanding of the educational concept (Novakovic 2012). At one end, there is a narrow understanding that focuses on developing the attitudes and skills that young people need to start and run their own business or become self-employed. At the other end, the emphasis is placed on entrepreneurship as a key competence whose goal is to empower young people and provide them with transversal skills for active citizenship, employability and possible, but not necessary, classic entrepreneurship and entrepreneurial activity within a larger organization. Entrepreneurial education was most comprehensively defined by the Executive Agency for Education, Audiovisual Activity and Culture at the European Commission. This agency's definition of entrepreneurial education is: "Entrepreneurship education refers to the development of students' skills and way of thinking, which enables them to turn creative ideas into entrepreneurial activity" (Education, Audiovisual and Culture Executive Agency 2016). This is a key competence for all students, encouraging personal development, active citizenship, social inclusion and employability. It is relevant at all levels of the lifelong learning process, in all educational disciplines and all forms of education and training (formal, informal and informal) that contribute to entrepreneurial spirit and action, with or without commercial purpose" (Education, Audiovisual and Culture Executive Agency 2016). In this way, we focus on developing entrepreneurial attitudes, skills and knowledge that could individually enable turning an idea into action, as well as that entrepreneurship does not only refer to economic activities and business creation but also to all other activities. The aim is to increase awareness of the mindset and skills needed to respond to opportunities, needs and challenges, such as problem solving, teamwork, communication, creativity, and resilience. The entrepreneurship education experience must have in itself the game theory and enable curiosity and imagination. It must be created in an environment that encourages experimentation and tolerates mistakes. This promotes "tolerant" behaviors, which also instill resilience and self-confidence. These skills can be taught and developed through tools such as games and simulation activities, where making mistakes has no consequences in the "real" world (Nagy 2020).

The impact of entrepreneurship education on intentions to start a business venture has been the focus of many studies and has found mixed results. The positive connection between entrepreneurship education and entrepreneurial intentions was discovered by many authors who confirmed that education is significantly correlated with the intention to create a new venture (Shah 2020). Also, entrepreneurship courses have a positive impact on people starting their businesses at some point in their careers. Entrepreneurship education increases people's interest in becoming entrepreneurs (Shah 2020). Bhatia and Levina (2023) stated that entrepreneurship is studied through formal higher education, both in basic studies and in the second cycle. Their research specifically singled out three of the best North American master's programs that have innovated curricula to meet the needs of today's students and have gone beyond conventional lectures. One of the approaches directs the lectures and work in the classroom to applications of real-life experience, while in the "operating room"the professor "simulates actual events"using a start-up example. This approach is formalized under the name "Creative Destruction Lab", similar to the "Endless Frontier Labs"program, which fosters Rotman's approach and focus on experiential learning. On the other hand, they singled out the Darden Business School of the University of Virginia, which directs education toward the recognition of entrepreneurship skills and risk acceptance by students. It is natural there is a "voice of reason"that directs people in terms of caution "what if this goes wrong", so young people should be encouraged to think and answer the question "what if this goes right"with a rational observation of risks. This higher education institution is the birthplace of effective entrepreneurship. This approach invites students to recognize their existing entrepreneurial resources and accept a certain amount of risk (effectuation.org 2023). This way of thinking contrasts with the more conventional business school approach that emphasizes risk minimization. In addition, while the business world is often known for its sharp, fiercely competitive nature, Darden's program instills in students a respect for the power of collaborative innovation by encouraging students to openly share ideas with their peers and use diverse insights and perspectives to co-create entrepreneurial endeavors (Harvard Business Review 2023).

7.4 Research Methodology

In order to answer the research question, which was defined in the very introduction of this chapter and to test the hypothesis, a 13-question survey was created and an online survey of companies from the Bosnia and Herzegovina was conducted. Anchoring was carried out in the period from December 2022 to July 2023 and a total of 40 responses were collected. Both private and public companies of Bosnia and Herzegovina participated in the survey.

The analysis of the collected responses and the testing of the hypothesis was performed in the IBM SPSS (*Statistical Package for Social Sciences*) 23 program. The hypothesis was tested using the chi-square test and contingency table analysis.

$$x^2 = \sum_{i=1}^{r}\sum_{j=1}^{c}\frac{\left(O_{ij} - E_{ij}\right)^2}{E_{ij}} \tag{7.1}$$

Source: Lovrič et al. (2006).

where

E_{ij} is expected frequencies

O_{ij} is observed frequencies

To eliminate missing data, the method of excluding missing data, exclusion in its entirety, from the group of traditional treatments of missing data was used. Exclusion of missing data in the whole (eng. *Listwise deletion*) implies deletion of observations on which at least one variable has missing data. This means that further analysis is performed on the basis of those observations that do not have any missing data (Mujkić and Poljašević 2021). The mentioned method was used because it was determined that it is a mechanism of missing data completely at random (MCAR), which implies that the probability of non-residual data on the observed variable is not related to other variables or to the values of the given variable Y. In order to confirm the above, Little's *was* also applied MCAR test. Based on the above, an overview of the analysis of missing data and replacement of missing data is given below.

Figure 7.1 shows that 2 out of 13, i.e., 15.38% of the questions contain missing data, i.e., answers. Missing data occur in 2 of the 40 observed companies in the sample, which is 5%.

Table 7.1 overviews the frequency and percentage of missing data for each analyzed question. On the basis of the table, it can be seen that the missing data, that is, the answers, appear on the 8th and 13th questions. One respondent did not answer the eighth question, and two respondents did not answer the thirteenth question.

Little's MCAR test is given in Table 7.2. Since it was obtained value smaller from level significance of 0.05, it is concluded that it is accepted zero hypothesis, which code Little's MCAR test reads: data completely by chance are missing. This means that the data is missing entirely by chance and that it is not present MAR (*missing at random*) or MNAR (*missing non at work*) mechanism missing data.

After analysis of missing data, we can replace them. In this chapter, as already mentioned, for the purposes of eliminating missing data, the method of excluding missing data was used, excluding it in its entirety from the group of traditional treatments of missing data.

The basis for using the mentioned method is reflected in the fact that by applying Little's MCAR test, it was determined that the data are missing completely by chance, that is, they are missing according to the MCAR mechanism. The results of applying the mentioned method are presented in Fig. 7.2.

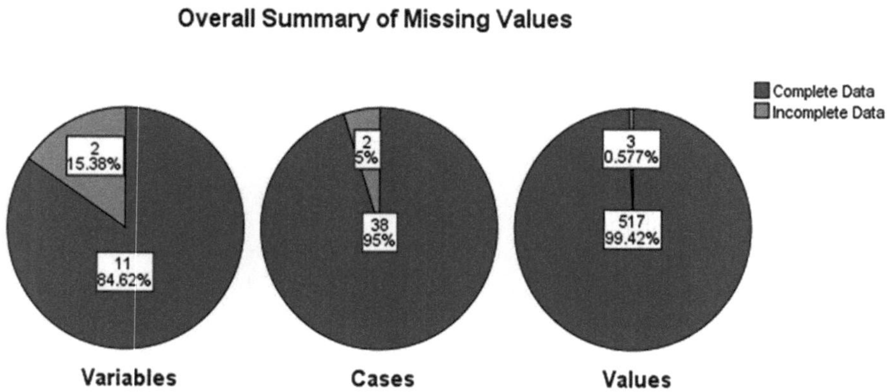

Fig. 7.1 Overall summary of missing values. Source: Authors calculation in SPSS program

Table 7.1 Overview of missing data by question

			Std.	Missing	
	N	I mean	Deviation	Count	Percent
1. Legal form of the company	40	2.10	.632	0	0.0
2. Ownership structure	40	1.88	.335	0	0.0
3. Activity of the company	40	6.85	3.577	0	0.0
4. Number of employees	40	118.18	188.805	0	0.0
5. Professional education of the company owner	40	2.10	1.317	0	0.0
6. Vocational education of the general director	40	1.98	1.250	0	0.0
7. Did your education provide you with the necessary knowledge for running a business	40	1.45	.504	0	0.0
8. Method of selecting candidates	**39**	**1.62**	**.493**	**1**	**2.5**
9. Are you satisfied with the qualifications of new employees when starting a job after higher education?	40	1.55	.597	0	0.0
10. Are you satisfied with the qualifications of new employees when starting employment after secondary education?	40	1.45	.504	0	0.0
11. What changes would you introduce in the education system?	40	2.98	1.672	0	0.0
12. In your opinion, what quality is lacking in young people?	40	8.85	5.036	0	0.0
13. In your opinion, is entrepreneurship a talent or can it be learned?	**38**	**1.21**	**.413**	**2**	**5.0**

Source: Authors calculation

Table 7.2 Little's MCAR test results

Question	Value
1. Legal form of the company	2.10
2. Ownership structure	1.88
3. Activity of the company	6.85
4. Number of employees	118.18
5. Professional education of the company owner	2.10
6. Vocational education of the general director	1.98
7. Did your education provide you with the necessary knowledge for running a business	1.45
8. Method of selecting candidates	1.61
9. Are you satisfied with the qualifications of new employees when starting a job after higher education?	1.55
10. Are you satisfied with the qualifications of new employees when starting employment after secondary education?	1.45
11. What changes would you introduce in the education system?	2.98
12. In your opinion, what quality is lacking in young people?	8.85
13. In your opinion, is entrepreneurship a talent or can it be learned?	1.21

Little's MCAR test: Chi-Square = 34.638, DF = 23, Sig. = 0.056
Source: Authors calculation

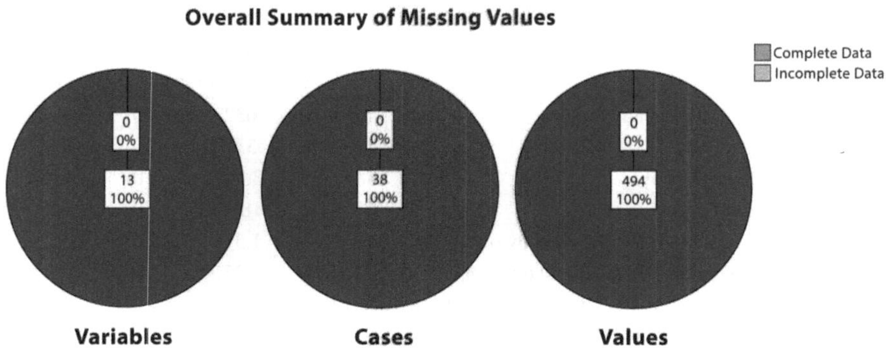

Fig. 7.2 Overview of missing answers. Source: Authors calculation

From the graph given in Fig. 7.2, it can be seen that all missing data, i.e., answers, are excluded from further analysis. Further analysis was performed on a sample of 38 companies.

The chi-square test is based on the 13th question, which states: "In your opinion, is entrepreneurship a talent or can it be learned?" The reason why the application of the chi-square test is based on this question is reflected in the fact that this question is directly related for the set hypothesis and research problem. Chi-square is applied with a risk of error of 5%.

7.5 Research Results

Out of 38 companies analyzed, 29, or 76.3%, are organized as limited liability companies, 10.5% as joint stock companies and other companies, and 2.6% as independent entrepreneurs. It is privately owned for approx. 87% of the surveyed companies, and the rest are public companies. The most significant percentage of surveyed companies is from information and communication activities 16%, followed by agriculture, forestry and hunting, and processing industry, 8% of surveyed companies each. Company owners most often have a first degree of higher education 53%, as do general managers 58%. Over half of the respondents, 53%, believe that their education provided the necessary knowledge to run a business. Candidate selection is done by the largest number of respondents through a competition, over 63%. Almost 53% of the respondents are satisfied with the qualifications of new employees when starting a job after higher education, and 58% are satisfied with the qualifications of new employees when starting a job after secondary education. 79% of respondents believe that entrepreneurship can be learned. The above results are presented in Table 7.3.

Half of the surveyed companies have less than 38 employees, and the other half have more than 38 employees; that is, the median number of employees in the surveyed companies is 38. The largest number of respondents believe that it is

Table 7.3 Overview of the answers from the questionnaire

		Frequency	Percent
1 Legal form of the company			
Valid	Joint stock company	4	10.5
	A limited liability company	29	76.3
	The rest	4	10.5
	Independent entrepreneur	1	2.6
	Total	38	100.0
2 Owner structure			
Valid	Public enterprise	5	13.2
	Private property	33	86.8
	Total	38	100.0
3 Activity of the company			
Valid	Construction	2	5.3
	Energy efficiency and renewable energy	1	2.6
	Plastic industry	1	2.6
	Information communication technologies	6	15.8
	The rest	11	28.9
	Agriculture, hunting, and forestry	3	7.9
	Wood processing and production of wood	1	2.6
	Manufacturing	3	7.9
	Food industry	2	5.3
	Transport, storage, and communications	1	2.6
	Professional, scientific, and innovative work	3	7.9
	Textile industry	1	2.6
	Wholesale and retail	2	5.3
	Financial intermediation	1	2.6
	Total	38	100.0
5 Professional preparation of the company owner			
Valid	And the level of higher education	20	52.6
	Completed II degree—Master of Science	5	13.2
	Completed III degree—Doctor of Science	2	5.3
	Middle school	11	28.9
	Total	38	100.0

(continued)

Table 7.3 (continued)

		Frequency	Percent
6 Professional preparation of the general director			
Valid	And the level of higher education	22	57.9
	Completed II degree—Master of Science	6	15.8
	Completed III degree—Doctor of Science	1	2.6
	Middle school	9	23.7
	Total	38	100.0
7 Did your education provide you with the necessary knowledge for running a business?			
Valid	That	20	52.6
	Not	18	47.4
	Total	38	100.0
8 Method of selecting candidates			
Valid	Based on a friend's recommendation	14	36.8
	Through a competition	24	63.2
	Total	38	100.0
9 Are you satisfied with the qualifications of new employees when starting a job after higher education?			
Valid	That	20	52.6
	Not	16	42.1
	There are no employed workers with VSS	2	5.3
	Total	38	100.0
10 Are you satisfied with the qualifications of new employees when starting a job after secondary education?			
Valid	That	22	57.9
	Not	16	42.1
	Total	38	100.0
13. In your opinion, is entrepreneurship a talent or can it be learned?			
Valid	It can be learned	30	78.9
	Talent	8	21.1
	Total	38	100.0

Source: Authors calculation

necessary for students and pupils to have more practical classes and work in companies 66%. Almost 16% of respondents believe that young people are not interested in work, while 13% believe that young people lack the ability to solve problems independently.

The null and alternative hypotheses are defined as follows:

H_0: Entrepreneurship are not skills and knowledge which can be acquired.
H_1: Entrepreneurship are knowledge and skills which can be acquired.

The calculated chi-square value statistics is 12.737, which can be seen from Table 7.4. Given that, the tabular chi-square value statistics smaller than calculated

Table 7.4 Chi-square test

Entrepreneurship is			
	Observed N	Expected N	Residual
It can be learned	30	19.0	11.0
Talent	8	19.0	−11.0
Total	38		

Test statistics	
	Opinion
Chi-Square	12.737[a]
df	1
Asymp. Sig.	.000

[a]0 cells (0.0%) have expected frequencies less than 5. The minimum expected cell frequency is 19.0
Source: Authors calculation

values, we reject zero hypothesis, and we conclude that entrepreneurship knowledge and skills which are meant to be acquired.

7.6 Conclusion

Summarizing previously stated, with risk errors of 5%, we can conclude that in Bosnia and Herzegovina, entrepreneurship is not observed as talent but as knowledge and skills that can be learned and acquired over time. What we need to point out are the limitations of the research. Namely, although the questionnaire is distributed to over 1000 addresses due to disinterested businessmen for the subject matter, Research 40 responses were collected by contacting businessmen daily by phone and reminding them to fill out the questionnaire.

Taking into consideration the research results and the importance of the educational sector, the goal is to create and nurture a culture of success, a culture that encourages young people to think about starting their own business venture. The job that falls to teachers and professors is the job of developing entrepreneurial skills through lectures, competitions, extracurricular activities, networking programs, encouraging competition, and the like. These are the factors that encourage young people to start a business venture. On the other hand, professors must continuously improve themselves, they must teach more affirmatively in order to follow the positive spirit of students and their desire to learn more, which makes lectures better and gives new value to students.

To conclude, although many entrepreneurs believe that passion for the idea and the desire for success are the two most essential qualities that adorn an entrepreneur (Hisrich et al. 2010), our research has shown that entrepreneurship can be learned.

Future research should be focused on the education sector. Mainly to guide how and what should be changed in a country's educational system so that young people are ready to start their business venture.

One of the recommendations that can be mentioned is that businessmen should help young people by creating a youth chamber of commerce. Experienced entrepreneurs with their knowledge and recommendations on what actions and mistakes their young colleagues should not make could help enthusiastic young people to create successful businesses (Novakovic 2012).

References

Aulet B (2013) Disciplined entrepreneurship. Wiley

Badulescu A, Badulescu D (2013) How entrepreneurial are doctoral students? Some evidence from Romania. J East Eur Res Bus Econ

Baletić Z (1982) Cantillon, Richard - Opća rasprava o prirodi trgovine. Centar za kulturnu djelatnost, Zagreb

Bhatia AK, Levina N (2023) Harvard Business Review. https://hbr.org/2020/08/can-entrepreneurship-be-taught-in-a-classroom?ab=hero-main-text

Brown C, Thornton M (2013) Entrepreneurship theory and the creation of economics: insights from Cantillon's Essai. Q J Austr Econ 401–419

Education, Audiovisual and Culture Executive Agency (2016) Entrepreneurship education at school in Europe. Evropska komisija/EACEA/Eurydice, Brussels

effectuation.org (2023). effectuation.org: www.effectuation.org

European Commission (2023) European Skills Agenda. https://ec.europa.eu/social/main.jsp?catId=1223&langId=en

European Union (2023) EUR-lex. https://eur-lex.europa.eu/EN/legal-content/summary/a-small-business-act-for-european-smes.html

Fridman M, Fridman P (1996) Sloboda izbora. Global Book, Novi Sad

Harvard Business Review (2023). https://hbr.org/2020/08/can-entrepreneurship-be-taught-in-a-classroom

Hisrich R, Peters M, Shepherd D (2010) Entrepreneurship eight edition—international edition. McGraw-Hill

Jussila J, Suominen A, Rainio T (2020) Entrepreneurship competence using educational hackathons in Finland. J Finnish Stud 23(2):32–73

Lackéus M (2015) Entrepreneurship in education: what, why, when, how. Center for Entrepreneurship, OECD, Paris

Lovrič M, Komić J, Stević S (2006) Statistička analiza - metodi i primjena. Ekonomski fakultet Banja Luka, Banja Luka

Mujkić E, Poljašević J (2021) Višestruka imputacija kao metod eliminacije nedostajućih podataka u SPSS-u. Financing 31–47

Nagy T (2020) Can entrepreneurship be taught? https://www.linkedin.com/pulse/can-entrepreneurship-taught-tobi-nagy/

Novakovic V (2012) Mogućnost razvoja omladinskog preduzetništva u Republici Srpskoj. EMC Rev Econ Mark Commun Rev 87–104

Novaković V (2018) Upravljanje inovacijama. Panevropski univerzitet Apeiron, Banja Luka

Novakovic V, Jakupovic S (2018) Omladinsko preduzetnistvo. Panevropski univerzitet Apeiron, Banja Luka

Novakovic V, Sašić B (2015) Odnos prosječne plate u javnom i privatnom sektoru u republici srpskoj shodno nivoima obrazovanja. EMC Rev Econ Mark Commun Rev 9(1):99–117

Popescu N (2013) Entrepreneurship—some general knowledge. Expert J Econ 28–32

Shah IA (2020) The moderating role of entrepreneurship education in shaping entrepreneurial intentions. J Econ Struct

World Economic Forum (2023) Future of jobs report 2023. World Economic Forum, Geneva. https://www.weforum.org/reports/the-future-of-jobs-report-2023/

Vesna Novaković is a Doctor of Science in management and business. She is an associate professor who has been teaching at the Faculty of Business Economics of the Pan-European University APEIRON, Banja Luk, since 2016. Appointed as Head of the Master's Study program, Entrepreneurial Management II cycle from 2021. From 2022, appointed as Head of the study program: Entrepreneurial Management. I specialize in internal financial controls, strategic financial planning, and program budgeting. She has a proven ability to prepare plans and programs for workshops on budgeting, financial planning, and the system of internal financial controls. She has special experience in teaching and cooperation with private and international organizations (UNDP, USAID, Sida). I have participated in numerous international conferences since 2011 and am the author of cited works in management, public finance, entrepreneurship, and innovation. She is the laureate of the Captain Miša Anastasijević Award for her professional contribution to the affirmation of entrepreneurial and social creativity in the Republic of Srpska.

Elvis Mujkić has a Master of Economic Sciences in accounting and auditing. He is the author and co-author of 9 articles published so far. He worked at Molson Coors BH d.o.o. from 2019 to January 2024. In the company, he worked in the finance department as a credit control analyst and financial planning and reporting specialist in the logistics department, where he worked as a business specialist and SAP expert. He also worked in the position of BIH procurement manager in the procurement department. He is currently employed as an accounting professor at the secondary school Srednja stručna i tehnička škola. He is a member of the Association of Economists of the Republic of Srpska SWOT. He also manages the Romanced II project, which the European Union and the Council of Europe implemented.

Dragan Milovanović is a Doctor of Science in the field of management and entrepreneurship. He is an Associate Professor, engaged in teaching at the Ekonomski fakultet Banja Luka, where he has been teaching for 15 years. He is the author and co-author of many scientific papers, published in national and international journals and he participated in international conferences in the fields of management. Engaged as a project manager of numerous projects.

Part IV
New Approach in CSR and Sustainability

Chapter 8
Grant Management with Special Focus on Allocated Funds to Non-Profit Organizations in Bosnia and Herzegovina

Vesna Novaković

Abstract The subject of this chapter includes the grant management in the Republic of Srpska and Federation Bosnia and Herzegovina, i.e., the method of allocating public financial resources to non-governmental sports organizations, with a special focus on football clubs and the performance of the allocated funds. The audit reports of the Supreme Audit Office of the Republic of Srpska Public Sector and the Federation of Bosnia and Herzegovina will be analyzed. The goal is to determine the extent to which grants awarded to non-governmental, i.e., sports organizations from public budgets in the Republic of Srpska and the Federation of Bosnia and Herzegovina are adequately managed. We define the problem with the question: Have competent institutions in grant management effectively implemented the principles of management according to results?

To answer the question, we will discuss the program budget concept and results-oriented planning. Following the above, the following hypotheses are tested: if grant planning meets the principles of results-oriented planning and if grants are awarded transparently and based on criteria aimed at achieving results.

Keywords Management · Public finance · Auditing · Internal control

V. Novaković (✉)
Faculty of Business Economy, Paneuroperan University Apeiron, Banja Luka, Bosnia and Herzegovina
e-mail: vesna.p.novakovic@apeiron-edu.eu

K. T. Çalıyurt (ed.), *New Approaches to CSR, Sustainability and Accountability, Volume VI*, Accounting, Finance, Sustainability, Governance & Fraud: Theory and Application, https://doi.org/10.1007/978-981-96-5937-1_8

8.1 Introduction

Numerous resolutions and conventions of the United Nations recognize the role of the civil sector in promoting the development of society, environmental protection and human rights, democracy, and good governance. The NGO sector includes a large number of people who are committed to these goals. The civil, i.e., non-governmental sector includes a wide range of organizations including all non-governmental, non-profit, independent organizations through which individuals or legal entities are organized in order to achieve common goals and ideals. Goals can be political, cultural, religious, ecological, social, economic, as well as goals related to health, socially neglected categories, and the like. In order to encourage the work of non-profit organizations in the Republic of Srpska and the Federation of Bosnia and Herzegovina, significant financial sums are allocated annually from public funds, entity budgets, and local communities for co-financing the projects of non-profit organizations. The financing of non-profit organizations is regulated by legal and by-laws that define the conditions and criteria that non-profit organizations should fulfill and satisfy in order to be allocated funds.

The subject of the analysis of this work includes the management of grants in the Republic of Srpska and the Federation of Bosnia and Herzegovina, that is, the method of allocating financial resources to non-governmental, sports organizations, with a special focus on football clubs, as well as the performance of the allocated funds. The audit reports of the Main Service for Auditing the Public Sector of the Republic of Srpska and the Federation of Bosnia and Herzegovina will be analyzed. The goal is to determine the extent to which grants awarded to non-governmental, i.e., sports organizations from public budgets in the Republic of Srpska and the Federation of Bosnia and Herzegovina are adequately managed. We define the work problem with the question: Have competent institutions in grant management effectively implemented the principles of management according to results? In order to answer the question, we will discuss both the concept of program budget and results-oriented planning. In accordance with the above, the following hypotheses are tested:

H1: Grant planning meets the principles of results-oriented planning.
H2: Grants are awarded transparently and based on criteria aimed at achieving results.

8.2 Grant Management Process, Legal Framework, and Concept

Grant management, in the broadest sense, includes a series of activities aimed at achieving the strategic goals and policies of the government (Cabinet Office, London 2023). In a narrower sense, grant management includes a series of activities related to: planning, awarding, monitoring, and evaluation of grant performance.

Grants (from the English word grant) are direct payments of a non-commercial nature to a specific user in order to implement certain measures or activities and fulfill pre-defined goals. Grants are a specific source of financing intended for certain institutions, that is, legal and natural persons who need an additional source of financing and which the grantor listed in the competition as potential beneficiaries of the grant in question. Grants are awarded for the purpose of financing the realization of a precisely defined project/program of the beneficiary.

Grants can be general or specific/dedicated and current and capital grants. Grants represent a significant source of financing, both in the initial stages of business, and in the later stages, with a clear purpose, purpose and goal. In this way, grant beneficiaries can increase their annual budgets and implement additional activities they would not be able to achieve from regular funds.

When it comes to the purpose of financing, it is mainly about budget funds that are allocated to non-profit organizations on three different bases: (a) as (grant) grants for financing their ongoing (regular) activities, (b) as grant funds for the implementation of certain projects, and (c) as grant funds for the organization of certain manifestations and events (Novakovic et al. 2024).

Grants represent a type of budget expenditure, and the same rules and regulations that govern the planning, adoption, and execution of the budget are applied to them (Novakovic et al. 2022). The basic idea underlying the principle of public expenditure is the demand for rational spending of public funds. In this sense, the Law on the Budget System of the Entities (Republic of Srpska and the Federation of Bosnia and Herzegovina) and the Law on Execution of the Entities' Budget for each fiscal year prescribe the adoption of spending programs, that is, the method of allocating funds from the positions of current and capital grants. In addition to the aforementioned laws, for the management of grants, laws that regulate certain areas or departments are also relevant. Relevant are the laws that prescribe the obligation to finance the subject areas from the budget in the form of grant funds. The form and content of the report on the expenditure of grant funds, the instructions for the preparation of periodic and annual reports, the period for which these obligations are drawn up, and the deadlines for their submission are regulated under the legal acts of the entity governments.

Institutions that manage grants are obliged to propose spending programs with allocation criteria for grant funds to entity governments.

Competent institutions that award grants are obliged, in accordance with the applicable legal and by-laws, to submit data on the beneficiary of the grants, the purpose of the funds, the amount of transferred funds, the amount of spent funds, and the amount of unspent funds on the prescribed form. Budget users are obliged to submit a written explanation of the intended use of funds along with this form, as well as the reason why the funds were not used. In the field of grants, in addition to the aforementioned laws, other laws, as well as by-laws such as decrees, regulations, decisions, instructions, orders, etc., are relevant. Grants are managed by relevant ministries and other budgetary institutions that are given the relevant authority by law or other acts.

Competent institutions dispose of budgetary funds for grants within their budget resources, define the objectives of granting grants, establish criteria for the selection of beneficiaries and the method of their evaluation, propose programs for spending funds to entity governments, adopt regulations and rules of procedure related to the selection of grant beneficiaries, implement procedures and procedures allocation of grants, monitor expenditure and evaluate the effects achieved by the allocation of funds, report to the competent ministries of finance and entity governments on the use of grants. For example, in the Republic of Srpska, the Law on the budget system of the Republic of Srpska (Ministry of Finance Government of the Republic of Srpska 2016), the Law on the implementation of the budget of the Republic of Srpska for a specific calendar year, by-laws passed by individual budgets and beneficiaries define the management of grants. Beneficiaries of grants should prepare programs and projects for which they apply for the use of funds from grants, implement programs and projects with financial support from grants, report to competent institutions on the implementation of programs and projects, and the use of financial resources.

Below are the steps of the process, as regulated by the current regulation. Each of the steps has its own specifics in relation to the requirements and principles of grant management according to results. In general, the management of grants is carried out in seven steps and three key phases: before the allocation of grant funds, the allocation of grant funds and the phase after the allocation. Each step consists of several activities depending on the angle of observation, whether it is a provider or a user of grant funds. This is shown in the following Fig. 8.1.

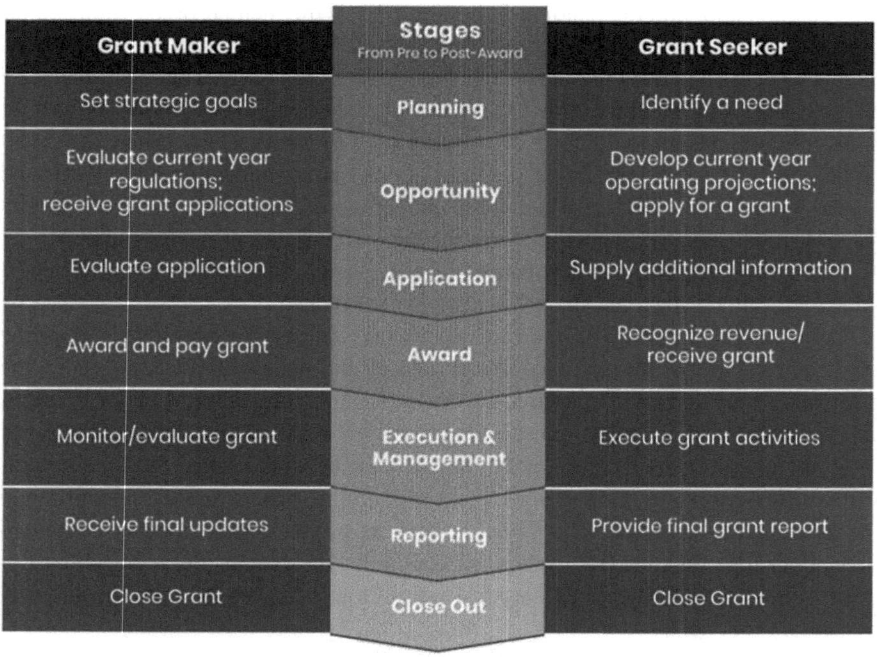

Fig. 8.1 Steps in the grant management process. Source: Amplifund (2024)

8.3 The Concept and Importance of the Program Budget

The budget is the primary instrument at the disposal of the government in the implementation of its policies. It represents a means by which strategic goals and commitments are realized through the implementation of services, programs, and activities aimed at meeting the social and economic needs of citizens. The program budget was introduced as an obligation of BiH in the process of European integration. The foundation of the program budget is results-based management.

To achieve budget quality, it is necessary to obtain information and performance facts as the tool used in determining budget allocations (OECD 2008). Performance-based budgeting is a public budgeting method that has been used by many countries as members of the OECD since the 1990s. According to the OECD report, performance-based budgeting has positive impacts on increasing the efficiency of public budgets (OECD 2008).

Budget management by results or program budget aims to improve the effectiveness and efficiency of public spending by connecting public sector financing with the results that are achieved. Through the program budget, information is a way to establish a clear connection between allocated budget funds and programs implemented by budget users, which should be aimed at achieving defined goals and government policies. Simply put, budgeting in a program format is a process in which funds are allocated to beneficiary programs, each of which contains clearly defined objectives and results that are planned to be achieved with the allocated funds. The program itself is created simply by grouping similar activities performed by the budget user, which have a common strategic and/or operational goal. Therefore, through the format of the program budget, the user is required to specify what is being done in order to achieve the planned goals and achieve the desired results and to show how much the achievement of the given goal will cost the budget. The program budget builds a system of values that enables budget users to better present how the funds will be used and the government and parliament to make better decisions about the priorities of distributing limited funds according to activities and results. Furthermore, budget planning based on the program format improves the transparency of resource allocation decisions. At the same time, the responsibility of the government and public sector managers for the effects and results of budget spending is also increasing. In other words, the focus is on implementing the requirements of the program budget system, which aims to systematically use performance information in budget decisions in order to improve spending prioritization and increase pressure for performance.

The foundation of the program budget is results-based management. Generally speaking, results-based management can be defined as a management strategy that implies that all relevant actors and processes directly or indirectly contribute to the achievement of the desired results. It is based on clearly defined responsibilities for results and requires continuous monitoring and self-assessment of progress, including performance reporting. In the public sector, focusing on performance results is a greater challenge than in the private sector, where profit is a clear indicator of performance. Results-based management is based on principles that are incorporated into all phases of management. Among the more important principles of results-based management are the following (UN Sustainable Development Group 2024):

- The grant spending program is based on previously determined goals and expected results
- Constant search for performance information
- Strengthening practices that encourage the use of performance information for the purpose of more effective management
- Good strategic and operational planning, including smart (SMART) definition of goals, achievement of results and continuous improvement process of performance information and
- The actual performance in the realization of goals is measured by objective performance indicators (UN Sustainable Development Group 2024)

Results-based management is based on a circular logical process in which formal performance information is used to establish cause-and-effect relationships between resources and activities, all with the goal of achieving the desired results. A logical framework (UNDP 2009) or chain of results is used to define these connections. The chain of results enables the development of interventions starting from inputs, through activities and outputs, and is finalized with results, impact, and feedback (City Colleges of Chicago 2021).

Activities of institutions can produce changes that are reflected in output, outcome, and impact. Outcomes and results play a central role in all performance budgeting models and are essential to any discussion of performance budgeting management. Evaluations show whether the undertaken activities influenced the results in the desired way, on the basis of which corrective measures are adopted. Results-based management emphasizes the importance of prioritizing expected performance through the use of objectives, desired results, and performance indicators (UN Sustainable Development Group 2024).

Program budgeting is part of the integrated budget process, which consists of:

- strategic and medium-term planning and the connection with the drafting of the framework budget document
- annual budget planning and definition of performance information
- execution, budget execution monitoring and reporting, and evaluation of implemented public policies

The key features of this integrated process are:

- Establishing strong links between long-term strategic planning, medium-term planning and the annual budget
- Monitoring and reporting on budget execution with reporting on the results of policy implementation measures

The process begins with long-term strategic planning and formulation of public policies. Strategic planning documentation usually consists of sector and/or ministry-level strategies, supplemented by an overall national development strategy and/or defined national priorities within the government's medium-term work program. The next stage is the preparation of the framework budget document. The outline budget document (DOB) is a preliminary draft of the budget for the coming year. DOB contains economic policy goals, macroeconomic estimates and projections, budget funds and expenditures projections, fiscal policy goals, and expenditure ceilings for the next year and the following two fiscal years, distributed by budget users and programs. Each budget user must comply with the DOB when preparing their annual request for budget funds.

The medium-term budget framework determines the main budget priorities of the government for the next three years because the framework connects defined policies (goals) and budget planning in the early phase of the budget cycle and includes policy changes.

The medium-term budget framework (MSF) is guided by the goals defined in the institution's strategic plan, and the budget is prepared based on that information. The link between the strategic plan and the medium-term budget framework enables the allocation of funds in accordance with priorities. Key features of the medium-term budget framework are as follows:

- enables operationalization of strategic plans of budget users and better definition of priorities
- obliges budget users to prepare a budget within the upper limits of expenditure in a three-year time frame
- directs budget users to budgeting toward performance based on achieved results and
- managers at all organizational levels must be involved in the budgeting process to ensure a greater degree of ownership and responsibility

8.4 Non-profit Organization, Concept, and Definition

When we look at the definition (UNDP 2014) of an association, it is stated that, unlike a company, the purpose of the establishment is not to make a profit, that is, to make a profit. The founders, members of the association, members of the management bodies, responsible persons, workers, or donors are not allowed to gain directly or indirectly profit or other material benefit from the activities of the association. This does not apply to compensation for work or reimbursement of expenses to those persons in connection with the realization of the legal goals and activities of the association that are regulated by the statute. Article 35 of the Law on Associations and Foundations of the Republic of Srpska defines that the property of an association or foundation consists of voluntary contributions and gifts that have a monetary value, financial resources from budget grants and taxes, subsidies and contracts with natural and legal entities, income from interest, dividends, profits from capital, rents, royalties and similar sources of passive income, income acquired through the achievement of goals and activities of the association or foundation; membership fee, when it comes to the association, and other income generated by the legal use of the organization's assets and funds.

The Law on Associations and Foundations of the Federation of Bosnia and Herzegovina in Article 35 defines: "Income of associations and foundations may include the following:

- membership fee in the case of an association;
- voluntary contributions and gifts from public institutions, natural and legal persons, both foreign and domestic, in cash, services or property of any kind;
- state subsidies or contracts with the state, public institutions, natural and legal persons, both domestic and foreign;
- income from interest, dividends, capital gains, rents, royalties and similar sources of passive income;
- income earned through the achievement of goals and activities of the association or foundation, as determined by the statute."

Those revenues can be used for the activities of the organization, as well as for the purchase of equipment, furniture, and other technical means for work. A registered association or foundation is obliged to properly keep business books, in accordance with accounting principles, and to prepare financial statements in accordance with the requirements set forth in applicable laws. Supervision over the legality and intended use and disposal of association and foundation funds is performed by their authorized body determined by statute and law, as well as competent authorities.

In accordance with the legal provisions, we distinguish between different sources of financing the work of non-governmental organizations. Some of them are sponsorship, donations, grants, procurement, compensation to non-profit organizations for services provided by them, public budgets (state, entity, cantonal, and municipal budgets).

Most often, funds received from donors are used for institutional support and for the implementation of programs; if funds are received from local self-government units, they can be intended for the implementation of project activities, then organizations can be allocated some space for use, so that this support is not expressed directly through the inflow of money, but as support in kind and the like (UNDP 2014).

8.5 Research Results

The source of the data used in our research are the reports of the main audit service (Supreme Audit Office of the Republic of Srpska 2024) of the Republika Srpska and the Federation of Bosnia and Herzegovina. We stayed at the level of the ministries of these entities. The reason for this approach is that due to the complicated organizational structure of Bosnia and Herzegovina, due to the additional level of cantonal authorities in the Federation of Bosnia and Herzegovina, the data would not be comparable. In Table 8.1, we provide an overview of the budget funds of the General

Secretariat of the Government of the Republic of Srpska that were allocated to sports and youth organizations for the period from 2019 to 2023.

Data provided in the Table 8.2 show the participation of grants to sports and youth organizations in the total grants of the budget user Ministry of Family, Youth

Table 8.1 Participation of grants to sports and youth organizations in total grants of the Secretariat General of the Government

Year	Name of the budget user	Budget BAM	Grants to non-profit entities in the country in BAM	Grants to sports and youth organizations	Share of grants in the total budget	Share of grants to sports and youth organizations in total grants
2019	Secretariat General of the Government	14,943,100	1,005,600.00	299,500.00	6.73	29.78
2020	Secretariat General of the Government	16,183,200	696,873.00	172,000.00	4.31	24.68
2021	Secretariat General of the Government	15,665,100	825,301.00	148,000.00	5.27	17.93
2022	Secretariat General of the Government	20,887,100	1,080,500.00	158,000.00	5.17	14.62

Source: Audit reports of the Supreme Audit Office of the Republika Srpska public sector

Table 8.2 Participation of grants to sports and youth organizations in the total grants of the budget user Ministry of Family, Youth and Sports in the Government of the Republic of Srpska

Year	Name of the budget user	Budget BAM	415200 current grants to sports organizations (current grants to sports and youth organizations and associations)	Share of current grants in the total budget
2019	Ministry of Family, Youth, and Sports	11,545,600	7,280,049.00	63.05
2020	Ministry of Family, Youth, and Sports	10,863,500	5,669,325.00	52.19
2021	Ministry of Family, Youth, and Sports	11,095,600	6,730,098.00	60.66
2022	Ministry of Family, Youth, and Sports	30,935,700	13,713,312.00	44.33

Source: Audit reports of the Supreme Audit Office of the Republika Srpska public sector

Table 8.3 Statements from the auditor's reports for the Ministry of Family, Youth, and Sports

Year	Name of the budget user	Budget BAM	
2019	Ministry of Family, Youth, and Sports	11,545,600	According to the conclusions of the Government of the Republic of Srpska, sports clubs were financed, namely: FK "Borac" Banja Luka (3,350,000 BAM), Handball club "Borac mtel" (170,000 BAM), FK "Radnik" Bijeljina (150,000 BAM), FK "Alfa" Modriča (120,000 BAM), FC "Zvijezda 09" (100,000 BAM), ŽRK "Borac" (100,000 BAM), ŽOK "Bimal-jedinstvo" (100,000 BAM), ŽKK "Orlovi" (100,000 BAM), ŽKK "Mladi Krajišnik" (100,000 BAM), then the amount of 50,000 BAM was awarded to: KK "Mladost" Mrkonjić Grad, KK "Leotar" Trebinje, KK "Bratunac", Boxing club "Slavija1961", then to an individual Goran Delic (30,000 BAM). Conclusions of the Government of the Republic of Serbia in progress 20,000 BAM each was awarded to the following clubs: FK "Tekstilac Kort", FK "Sloga", FK "Slavija", FK "Rudar", FK "Podrinje", FK "Krupa", FK "Kozara", FK "Jedinstvo" FK "Zeljezničar", university volleyball club "Bl volley", student club "Student-igokea", swimming club "Delfin", youth club "Slavija 1996", youth club "Igman", youth club "Drina princip", KK "Heo", KK "Sutjeska", KK "Sokolac", KK "Sloboda 73", KK "Radnik-bn basket", KK "Prijedor", KK "Varda-he", KK "Budućnost bn", KK "Borac", ŽKK "Sloboda", ŽKK "Rudar sport", ŽKK "Leotar 03", water polo club "Banja luka".
2020	Ministry of Family, Youth, and Sports	10,863,500	The current grant for top sports was awarded according to the requirements for co-financing of regular and competitive activities and according to the Conclusion of the Government of the Republic of Srpska. By the conclusion of the Government of the Republic of Srpska, financial assistance was approved to the Football Club "Borac" in the amount of 1,000,000 BAM. Current grants for the financing of sports clubs and sports events in the Brčko District are reported in the amount of 168,000 BAM, and are approved based on the Rulebook on conditions and criteria for financing the carriers of the development of sports in the Republic of Srpska. It was approved according to the competition (50,000 BAM), through the current request (18,000 BAM) and according to the Conclusion of the Government of the Republic of Srpska to the Football Club "Jedinstvo" (100,000 BAM). Evaluation of applications for the competition by the Commission for the Main Service for Auditing the Public Sector of the Republic of Srpska Report on the financial audit of the Ministry of Family, Youth and Sports for the period 01.01-31.12.2020. year 15, the assessment and selection of projects is not in accordance with the criteria defined in Article 11 of the aforementioned rulebook, which stipulates that the assessment is carried out in accordance with 13 set criteria.

(continued)

Table 8.3 (continued)

Year	Name of the budget user	Budget BAM	
2021	Ministry of Family, Youth, and Sports	11,095,600	Current grants to sports organizations were awarded through current requests (1,412,265 BAM), according to the Sports Development Program (270,033 BAM) and according to the Conclusion of the Government of the Republic of Srpska. The Government of Republika Srpska made a decision to provide 40,000 BAM of financial aid to the Football Club "Alfa" Modrič.
2022	Ministry of Family, Youth, and Sports	30,935,700	Whitin the Auditor's Report for the Ministry of Family, Youth, and Sports in 2022. were no special explanations regarding the grants

Source: Audit reports of the Supreme Audit Office of the Republika Srpska public sector

and Sports in the Government of the Republic of Srpska. From Table 8.3, we can see that in addition to the competition procedure, according to the Decree of the Government of the Republic of Srpska, a staggering amount of funds was allocated to sports organizations, with a special emphasis on football clubs.

While in Tables 8.4 and 8.5, we can see the recommendation of the main public sector audit service. They stated that the projects of sports organizations were not evaluated in accordance with the procedure nor was there an adequate monitoring of the use of allocated funds.

In 2024, the Audit Office (Audit Office of the Institutions in the federation of Bosnia and Herzegovina 2024) published the Performance Audit Report on the topic "Management of grants from the Budget of the Federation of Bosnia and Herzegovina according to results." The aim of the audit was to examine whether the competent institutions had undertaken activities so that the management of grants from the FBiH Budget was directed toward results. The audit determined:

- In the planning phase, the institutions did not conduct comprehensive and documented analyses of the results achieved in the previous cycles of granting grants, nor did they conduct analyses of the situation and needs in the area. These analyses are necessary in order to produce quality input information necessary for the preparation of planning documents for the awarding of grants. Although planning is a fundamental phase in results-based grant management, institutions have not paid enough attention to it. The goals are general and most often defined as regular activities under the jurisdiction of the observed institutions. Not a single institution used information from documented analyses to (re)define measurable goals and expected results. These are cases where it is not possible to clearly see from the planning documents which goals the institution wants to achieve in a given year by awarding grants.
- No institution was able to ensure that all potential beneficiaries of grants have equal access to all essential information related to the use of grants, nor were the names of the selected beneficiaries and the amounts of grants awarded available

Table 8.4 Recommendation for the Ministry of Family, Youth, and Sports

Year	Name of the budget user	Recommendation
2019	Ministry of Family, Youth, and Sports	It is recommended to the minister to ensure that when granting grants, project evaluation is carried out as well as supervision of grants awarded to sports organizations, in accordance with Article 7 and 11 of the Rulebook on conditions and criteria for financing sports development bodies in the Republic of Srpska.
2020	Ministry of Family, Youth, and Sports	4) when awarding grants, it evaluates projects as well as supervises grants awarded to sports organizations, in accordance with Articles 7 and 11 of the Rulebook on Conditions and Criteria for Financing Sports Development Providers in the Republic of Srpska
2022	Ministry of Family, Youth, and Sports	In accordance with the Rulebook on special criteria and procedure for the distribution of funds from games of chance ("Official Gazette of the Republic of Srpska", number 120/20), funds were distributed in the amount of 1,004,500 BAM, namely in the field of sports (604,500 BAM).

Source: Audit reports of the Supreme Audit Office of the Republika Srpska public sector

to the public. The selection of grant beneficiaries was made on the basis of a public call, which at most institutions had different content. The allocation of grants, which was carried out without a public call, puts potential beneficiaries in an unequal position.

- The application of existing criteria for the selection of beneficiaries does not ensure the selection of quality projects/programs, the implementation of which contributes to the achievement of established goals. One of the reasons is that the criteria are not concrete and measurable enough, which is a consequence of deficiencies in the planning phase.
- The institutions did not use the opportunity to include in the contracts on the use of grants the requirements that grant beneficiaries submit quality information and evidence about the results and effects achieved by the implementation of projects/programs financed from grant funds. The above information represents inputs necessary for performance evaluation and reporting to the FBiH government.
- The monitoring system does not provide the quality information needed for results-oriented reporting. The institutions did not create basic assumptions in order to use the existing monitoring system for reporting not only on the intended expenditure of grants but also on their output and final results, which is necessary for the management of grants according to the results (Table 8.6).

Performance is the achievement level description of the activity/program/organizational policy implementation in realizing the strategic goals set by the organization, customer satisfaction, and its contribution to society's economic development (Bernardin et al. 1995). From the various literatures related to the definition of performance-based budgeting, there is a universal definition of performance-based budgeting which is the budget needed to finance expected activities (the outputs and

Table 8.5 Recommendation for the Secretariat General of the Government

Year	Name of the budget user	Recommendation
2019	Secretariat General of the Government	It is recommended to the general secretary to ensure that: (1) the approval of the Government of the Republic of Srpska is obtained for the plan for spending subsidies and grants in accordance with the Law on the execution of the budget of the Republic of Srpska for each fiscal year, and that in the event of a change in the purpose of the given funds, approval is obtained in accordance with according to Article 8 of the Rulebook on the Allocation of Financial Aid and the Distribution of Budget Reserve Funds, (2) together with the requests for the allocation of grant funds and financial aid, submit the documentation prescribed by Article 4 of the Rulebook on the Allocation of Financial Aid and the Distribution of Budget Reserve Funds
2020	Secretariat General of the Government	The recommendation given for grants has been implemented, while the recommendation regarding obtaining the consent of the Government of the Republic of Srpska to the Plan for spending subsidies and grants has been partially implemented. During 2020, the consent of the Government of the Republic of Srpska was subsequently obtained for part of the subsidies granted, which is not in accordance with Article 5, paragraph (3) of the Law on the Execution of the Budget of the Republic of Srpska for 2020 ("Official Gazette of the Republic of Srpska", number 112/ 19, 60/20 and 122/20) which prescribes that the use of budget funds for subsidies and grants can be made upon obtaining the approval of the Government on the Spending Plan of those funds.
2021	Secretariat General of the Government	As stated under point 6.1.2.1. reports: Reports on the expenditure of funds were not submitted for part of the granted grants, which is not in accordance with Article 14, paragraph (4) of the Law on Budget Execution of the Republic of Srpska. Audit tests included the amount of 271,000 BAM of granted grants (for 17 beneficiaries). For the granted grants in the amount of 75,000 BAM, reports on the expenditure of funds were not submitted, which is not in accordance with Article 14, paragraph (4) of the Law on the Execution of the Budget of the Republic of Srpska for the year 2021, which stipulates that the beneficiaries of funds are budget reserves and funds for grants obliged to submit reports to donors on the expenditure of received funds no later than the deadline for submitting annual financial reports, as well as Article 8 of the internal Rulebook on sports and youth organizations and associations (148,000 BAM). It is recommended to the general secretary to ensure that reports on the expenditure of the allocated grant funds are submitted in accordance with Article 14, paragraph (4) of the Law on budget execution of the year in question.

Source: Audit reports of the Supreme Audit Office of the Republika Srpska public Sector

outcomes are set out in performance targets) so each fund spent can be measured efficiently and effectively (Martí 2013).

Performance-based budgeting itself is a budgeting connecting each funding included in the expenditure (output) with the results achieved (outcome) such as efficiency, effectiveness, and economy as an instrument to achieve program goals and objectives (Duncan P Last and Marc Robinson 2009).

Table 8.6 Recommendation for the grant management in the Federation of Bosnia and Herzegovina

Ensure high-quality input information necessary for the development of planning documents for results-oriented grant awarding.
Establish specific and measurable goals and expected results for each grant cycle.
Ensure transparency of grant allocation
Determine the criteria for the selection of grant beneficiaries and the method of their evaluation, which will ensure the selection of a project/program, the implementation of which contributes to the achievement of the established goals.
Oblige grant beneficiaries to submit quality information and evidence about the results and effects achieved through the implementation of projects/programs.
Improve the monitoring system in such a way that it ensures quality feedback on the results of project/program implementation.
Establish the practice of performing essential and comprehensive analyses of the results and effects achieved by the implementation of projects/programs.
Ensure regular reporting to the FBiH Government on the effects of grants on the achievement of strategic goals and commitments

Source: Audit Office of the Institutions in the Federation of Bosnia and Herzegovina

It aims to improve the quality of public budgets. Based on it, the researchers argue that the quality of public budgets is achieved when (1) state expenditures are carried out efficiently and effectively (operational efficiency), (2) public financial accountability increases, and (3) transparency is achieved in public budget management.

Based on all of the above, it is evident through the tables that in the Republic of Serpska and Federation of Bosnia and Herzegovina, there is still no adequate system for managing grants based on results. Although there is an obligation to introduce a program budget and distribute budget funds in accordance with the achieved results, this is still not confirmed in practice.

8.6 Conclusion

Based on the collected and processed data from the audit reports, it is evident that on an annual level, significant funds are allocated from the budget of the Republika Srpska and the Federation of BiH for the financing of non-profit sports organizations, whereby the Ministry of Family, Youth and Sports, and the General Secretariat of the Government of the Republic of Sprka stand out as the most significant budget users according to the amount of allocated funds. Funds, which are awarded as non-refundable non-profit organizations, are most often used to finance their ongoing and regular activities, as grant funds for the realization of certain projects and as grant funds for the organization of certain manifestations and events. Sports clubs and organizations lead the way in terms of participation in financing.

The analysis showed that the legal framework clearly set and defined the rules of conduct regarding the rules and procedures for allocating funds to non-profit organizations. However, in practice, it happens that the institutions that allocate funds do not comply with the legal provisions in question, because funds are allocated to non-profit organizations upon request, without conducting a tender procedure and announcing a public call. The allocation of funds to non-profit organizations is accompanied by low transparency and the absence of basic information about the purpose of the funds, the activities carried out, the results achieved, and the expenditure itself. That is, the users of the funds do not submit reports on the expenditure of the received funds.

Based on all of the above, we can conclude that both hypotheses have been refuted. In the Republic of Srpska and the Federation of Bosnia and Herzegovina, grant planning does not meet the principles of results-oriented planning. The dry hypothesis was also refuted because we saw through the analysis of audit reports that grants are not awarded transparently and based on criteria aimed at achieving results.

Also, we can conclude that there is a clearly defined legal regulation, that is, the legal framework has set norms of behavior regarding the allocation of public financial resources to non-profit organizations. However, the lack that is evident is the application of those legal solutions in practice. Although the allocation of financial resources to non-profit organizations is carried out through a public call, there is still a lack of transparency in terms of the purpose and effects of the allocated funds.

The recommended recommendation is to introduce a system of prior controls regarding the assessment of the capabilities of non-governmental organizations (acquisition of necessary equipment, expertise of personnel engaged in the implementation of project activities, etc.) in order to further reduce the risk of misuse of allocated funds. From the aspect of control activities, both local and entity, the authorities should continuously monitor the realized effects of allocated funds in relation to the objectives of public calls, according to which non-profit organizations were allocated funds.

The aforementioned measures would additionally affect the transparency, efficiency, and effectiveness of spending public funds allocated to non-profit organizations.

References

Amplifund (2024). https://www.amplifund.com/blog/grant-management. Accessed 17 Apr 2024

Audit Office of the Institutions in the Federation of Bosnia and Herzegovina (2024) Grant management of the Federation of Bosnia and Herzegovina Budget. Audit Office of the Institutions in the Federation of Bosnia and Herzegovina, Saraejvo

Bernardin HJ, Kane JS, Ross S, Spina JD, Johnson DL (1995) Performance appraisal design, development, and implementation. In: Handbook of human resource management. Kogan Page, pp 462–493

Cabinet Office, London (2023) Government_Strategy_for_Grants_Management_AW. September. https://assets.publishing.service.gov.uk/media/65169fa77c2c4a001395e23f/3335_Government_Strategy_for_Grants_Management_AW.pdf. Accessed 22 Apr 2024

City Colleges of Chicago (2021) Grants management manual. City Colleges of Chicago, Chicago
Martí C (2013) Performance budgeting and accrual budgeting: a study of the United Kingdom, Australia, and New Zealand. Public Perform Manag Rev 37(*1*):33–58
Ministry of Finance Government of the Republic of Srpska (2016) Law on the budget system of the Republic of Srpska. Official Gazette of the Republic of Srpska, Banja Luka
Novakovic V, Aleksic M, Milovanovic M (2022) Own revenues in the public sector. Časopis za ekonomiju i tržišne komunikacije/Economy and Market Communication Review 12:126–135
Novakovic V, Jokic M, Milovanovic M (2024) Impact. EMC 2024:50–62
OECD (2008) Performance budgeting: a users' guide. Organisation for Economic Co-Operation and Development
Duncan P Last & Marc Robinson (2009). A Basic Model of Performance-Based Budgeting, IMF Technical Notes and Manuals 2009/001, International Monetary Fund. https://ideas.repec.org/p/imf/imftnm/2009-001.html
Supreme Audit Office of the Republic of Srpska (2024) Audit reports. Supreme Audit Office of the Republic of Srpska, Banja Luka
UN Sustainable Development Group (2024) UN Sustainable Development Group. https://unsdg.un.org/resources/unsdg-results-based-management-handbook. Accessed 17 Apr 2024
UNDP (2009) PME handbook. UNDP. http://web.undp.org/evaluation/handbook/documents/english/pme-handbook.pdf. Accessed 17 Apr 2024
UNDP (2014) Handbook for non-governmental organizations in Bosnia and Herzegovina. UNDP

Vesna Novaković is a Doctor of Science in management and business. She is an associate professor who has been teaching at the Faculty of Business Economics of the Pan-European University APEIRON, Banja Luk, since 2016. Appointed as Head of the Master's Study program, Entrepreneurial Management II cycle from 2021. From 2022, appointed as Head of the study program: Entrepreneurial Management. I specialize in internal financial controls, strategic financial planning, and program budgeting. She has a proven ability to prepare plans and programs for workshops on budgeting, financial planning, and the system of internal financial controls. She has special experience in teaching and cooperation with private and international organizations (UNDP, USAID, Sida). I have participated in numerous international conferences since 2011 and am the author of cited works in management, public finance, entrepreneurship, and innovation. She is the laureate of the Captain Miša Anastasijević Award for her professional contribution to the affirmation of entrepreneurial and social creativity in the Republic of Srpska.

Chapter 9
A Conceptual Study on Sustainable Consumption

Aysun Duruş and Şeniz Özhan

Abstract The aim of this study is to address the concept of sustainable consumption, which has become one of the most important issues of recent times in our world, where the global environmental crisis is increasing, from the consumer perspective. Unsustainable production and consumption are one of the biggest threats to humanity and future generations. Sustainable consumption argues that we should use resources to meet our basic needs and not use more resources than we need. Limited resources despite unlimited consumption make the concept of sustainability important. Sustainability is considered a participatory process that creates a social perspective by ensuring the prudent use of all social, scientific, natural, and human resources in society. In order to manage resources consciously within the framework of sustainability, the concepts of reduce, reuse and recycle, summarized as 3-R, have been mentioned in order for both producers and consumers to act together. Sustainability, which refers to a process, includes both the present generation and future generations to have better living conditions. It is very important to understand the factors that determine consumer behavior in creating sustainable consumption awareness and all economic, social, and environmental stakeholders at

Prepared from Aysun Duruş's master's thesis entitled "Analysis of consumer perspective sustainable consumption studies with text mining", under the supervision of Assoc. Prof. Dr. Şeniz ÖZHAN.

A. Duruş
Tekirdağ Namık Kemal University, Tekirdağ, Türkiye
e-mail: adurus@nku.edu.tr

Ş. Özhan (✉)
Faculty of Economics and Business Administration, Department of Business Administration, Tekirdağ Namık Kemal University, Tekirdağ, Türkiye
e-mail: serdem@nku.edu.tr

K. T. Çalıyurt (ed.), *New Approaches to CSR, Sustainability and Accountability, Volume VI*, Accounting, Finance, Sustainability, Governance & Fraud: Theory and Application, https://doi.org/10.1007/978-981-96-5937-1_9

national, regional, and international levels should be involved in the change process for sustainability. The Ecological Footprint, which is widely used by scientists, businesses, governments, individuals, and institutions to monitor the ecological situation and to advance resource use and sustainable development, is considered as an indicator of total population and consumption status.

Keywords Sustainable consumption · Sustainability · Consumer

9.1 Introduction

Sustainability is one of the remedies for the ecological crisis, which is the most important crisis our world has faced in the early twenty-first century. Therefore, the concept of sustainability has been the subject of much debate in Western literature since the early 1990s. The emergence of the concept of sustainability, which aims to minimize the negative effects of the fact that humans have been consuming their environment since the beginning of their existence, is associated with the ecological crisis and the Green Movement in the 1970s. Sustainability, which came to the agenda as a remedy for the ecological crisis, promises a sustainable future.

According to the Living Planet report published by the World-Wide Fund for Nature (WWF), an international non-governmental organization aiming to prevent damage to nature, "Our planet needs 1.5 years to replenish the natural resources consumed in 1 year. If we continue as we are today, even two planets will not be enough for us in 2030!", emphasizing the extent of the negative impact of ecological footprints on the ecosystem. People on Earth are multiplying exponentially in numerical terms. According to the 2021 World Population Data Sheet (PRB), the world population first reached 1 billion in 1800. A century later, it reached 1.7 billion in 1900, 2.5 billion in 1950, and over 7 billion in 2020. By the middle of this century, population growth is expected to continue and exceed 9 billion. Reducing the "ecological footprint", which is defined as the impact of a human being on the ecosystem and natural resources while living, will enable us to leave a habitable planet for future generations (Taluğ & Birden, 2014, p. 79).

In this study on sustainable consumption, the consumption behaviors of individuals for the continuity of our planet and therefore human life are discussed from the perspective of sustainability, and it is aimed at raising awareness by revealing the determinants of sustainable consumption. In the introduction part of the study, the importance of sustainability is mentioned, the concept of sustainable consumption is discussed in various aspects in the second part and an overview is given in the conclusion part.

9.2 Sustainable Consumption

9.2.1 The Concept of Sustainability

In the meeting organized by the United Nations World Commission on Environment and Development (Brundtlant) in 1987, the concept of sustainability was mentioned as follows; "Humanity has the ability to make development sustainable to meet the needs of the present without compromising the ability of future generations to meet their own needs" (Report of the World Commission on Environment and Development, 1987). Sustainability is to ensure that the activities carried out in the form of using renewable resources instead of non-renewable resources by improving natural resources through environmental regulations and human behavior do not pose a threat to the life of the world in the long term (Marın & Yıldırım, 2004, p. 253). Sustainability is to protect and improve the existing environmental quality, to protect human health, to ensure the proper use of natural resources, and to evaluate all environmental problems at the international level (Keleş & Hamamcı, 1993, p. 50). Societies that think consumption-oriented instead of production-oriented think in the consumption economy, in addition to giving movement to the economies, damage the world in an irreversible way (Ergen, 2014, p. 90).

Sustainability can include three conceptual dimensions: economic, ecological, and social. The implementation of fair wage policies for consumers and producers on a sectoral basis reflects the economic dimension of the concept of sustainability, while sensitivity to the natural environment reflects the ecological dimension. The harmonization of production stages with the priorities and needs of society involves the last dimension of the concept, the social dimension. Another important point for society is the support of country administrators for sustainable development policies (Vermeir & Verbeke, 2008, p. 542). Limited resources, despite unlimited consumption make the concepts of sustainability and sustainable development important. Sustainability is a participatory process that creates a social perspective by ensuring the prudent use of all of society's social, scientific, natural, and human resources (Gladwin et al., 1995, p. 874).

9.2.2 Sustainable Consumption Concept and Emergence Process

The terminology of sustainable consumption dates back to the 1970s. In 1972, the Limits to Growth Report of the Club of Rome, a think tank that describes itself as "a group of world citizens sharing common knowledge for the future of humanity", drew attention to the impact of increasing levels of prosperity in terms of resource scarcity and environmental degradation. The oil price spikes of 1973/4 and 1979/80 showed that consumption levels could not continue as usual. Following the shocks, oil prices fell and the predictions of resource scarcity did not materialize. But the relevance of consumption patterns to major environmental issues such as climate change, ozone depletion, and hazardous waste management became increasingly apparent in the early 1990s. The late 1990s saw a major increase in initiatives for sustainable consumption. The United Nations Environment Programme (UNEP) launched a sustainable consumption network, integrating sustainable consumption policies into the Consumer Protection Principles, and in 2001 published a strategic document highlighting the opportunities provided by the new sustainable consumption focus (UNEP, 2001). In 2002, the WSSD (World Summit on Sustainable Development) report identified "changing consumption and production patterns" as one of the three overarching goals for sustainable development.

The focus of sustainable consumption is on economic activities such as the selection, use, and disposal of goods and services and how these can be changed to provide social and environmental benefits. Sustainable consumption argues that we should use resources to meet our basic needs and not use more resources than we need (IIED, 1998). Consumption behaviors have many environmental consequences, such as CO2 emissions, which cause global warming due to biodiversity loss and depleting natural resources. Recently, in the examination of these effects, there have been factors such as the rapid increase in population and economic developments. Some of these factors (Karalar & Kiracı, 2011, p. 63):

- Globalization and the economies of scale it causes.
- Falling price levels of products.
- Cultural changes such as urbanization and women becoming more active in business life.
- Improved quality of life for individuals and.
- The emergence of environmental problems.

The concept of sustainable development, which aims to protect our environment along with economic development, has led to sustainable consumption. Sustainable consumption is a decision-making process that emphasizes the social responsibility of the consumer as well as individual needs and desires (Vermeir & Verbeke, 2006, p. 169). When an individual decides whether or not to purchase a product or service, he/she has the potential to contribute to a sustainable consumption model, albeit to a lesser extent (Young et al., 2010, p. 18).

The concept of sustainable consumption was first brought to the agenda within the scope of the Sustainable Development Action Plan of the Earth Summit held in Rio in 1992 (Bayazıt Hayta, 2009, p. 145). According to the OECD reports, where the definition of the concept was clarified, sustainable consumption is defined as "the use of goods and services that offer a quality life by meeting basic needs while minimizing the use of natural resources, toxic substances, waste emissions and environmentally harmful substances from a life cycle perspective, taking into account the needs of future generations" (OECD, 2002). Sustainable consumption draws attention to the concepts of avoiding consumption, attitudes, awareness, and creating alternatives rather than minimizing individual and social consumption. Various alternative approaches have been produced in order to realize the phenomenon of sustainable consumption, which is a form of consumption that includes methods that minimize the damage to natural life based on the use of natural resources within the scope of sustainability. The first of the two alternative approaches developed for this purpose is to reduce the level of consumption by engaging in consumption behaviors that are more aware of environmental protection. In this way, it is essential to transform consumption behaviors in daily life into a more sustainable way (Karalar & Kiracı, 2011, p. 63). The second alternative approach proposed in this regard is to both transform consumer behavior and produce more sustainable products and services. Five factors suggested by Seyfang (2006) for the realization of sustainable consumption:

- Use of local products
- Re-incorporating post-consumer waste back into the production process
- Creating an active and thriving society by developing sustainable communities
- Establishing institutional organizations to solve consumption problems and
- Directing the society toward environmental behaviors

Sustainable consumption aims to improve the consumption of goods and services that have a positive impact on consumers and the environment by combining social, economic, cultural, or political practices at the level of individuals, households, businesses, and communities (Niva et al., 2014, p. 465).

9.2.3 Dimensions of Sustainable Consumption

The dimensions of the concept of sustainable consumption, the 3Rs—reduce, reuse, and recycle—were first used by former Japanese Prime Minister Junichiro Koizumi at the G8 summit in 2004 (Visvanathan & Kumar, 2007, p. 11). In a study conducted in Mexico on the dimensions of reuse and recycling, it was found that skills were more determinant in actual behaviors and thoughts were more determinant in desired behaviors (Corral-Verdugo, 1997, p. 135). Li (2003), in a study conducted in China, determined that gender, age, and household income are three effective factors in exhibiting sustainable consumption behavior due to recycling. The study revealed that older women and women with lower household income levels are more likely to exhibit sustainable consumption behavior due to recycling. In order to manage resources consciously within the framework of sustainability, 3-R concepts are mentioned in terms of both producers and consumers acting together. These are (Altıok & Babaoğul, 2008, cited in Kayıkçı et al., 2019, p. 81):

- **Reduce:** It means turning to products that are used in a way that takes care to reduce the amount of waste after the production and consumption of products.
- **Reuse:** It means the reuse of some or all parts of products.
- **Recycle:** Obtaining a new product from waste is the process of transforming products. When the principle of first reduce and then recycle by reusing is considered in the form of an inverted pyramid, reuse will be needed at the rate of waste reduction and recycling will be required at the rate of waste reuse.

The concept of sustainable consumption is to prefer environmentally friendly products that do not harm the environment or have minimal damage to the environment by reducing the amount of consumption while meeting consumer needs (Hansen & Schrader, 1997, p. 443). As the concept of environment has gained more importance for consumers as a result of the damage caused to the environment by modern consumption patterns, consumers have shifted from traditional products to environmentally friendly products. When purchasing environmentally friendly products, it is thought that consumers will act more decisively if they have environmental values, have environmental purchasing experiences, have received environmental education, and have access to environmentally friendly products (Gürbüz & Ayar, 2018, p. 920).

In the report "Towards Sustainable Household Consumption?" prepared by the OECD in 2002, general principles were established to promote sustainable consumption. Criteria such as recycling, life cycle, use of eco-products have been put into practice. Effective mechanisms were created to promote sustainable consumption, and it was aimed for governments to support sustainable consumption policies, and sustainable consumption was analyzed in three dimensions: economic, social, and environmental (OECD, 2002). Özgül (2010) mentioned two dimensions of sustainable consumption, namely, saving tendency and environmental orientation, and argued that saving tendency is a more preferred sustainable consumption behavior than direct environmental protection. In another study, it was determined that while women were more careful in behaviors related to the use of money resources such as turning off the lights in unused rooms, fully filling and running the dishwasher and washing machine, turning off the television when not watching, and paying attention not to overheat the house (23 °C and above) in winter, they were less careful in behaviors directly related to environmental protection such as collecting recycling products such as paper and glass separately from others, paying attention to environmental friendliness when purchasing cleaning and cosmetic products, and preferring products with recyclable packaging (Şener & Hazer, 2008, p. 291).

9.2.4 Stakeholders of Sustainable Consumption

The problem of how to guide individuals' sustainable consumption behavior is an important issue for both policymakers and researchers. What is needed is to determine how sustainable consumption behaviors affect natural and social systems, as well as how these behaviors can be made more sustainable (Geiger et al., 2018, p. 2). In the industrial sector, sustainable consumption refers to the most efficient use of technology and resources, producing more environmentally friendly and economical products, and informing consumers about advertising, market, and product issues. However, the incentives required for these to be realized need to be identified. All stakeholders in sustainable consumption should cooperate in developing viable strategies that can solve the problem of today's consumption trends (Kong et al., 2002, p. 109). The international literature lists various recommendations for sustainable consumption in the context of producers and consumers (McLaren, 2007, p. 5).

To produce sustainable products and services:

- Development of new environmental technologies and eco-innovation
- Development of cleaner production processes and eco-efficient products and services
- Promoting environmental technologies and eco-innovations (including the collection and dissemination of information) by businesses (and especially in developing countries)
- Voluntary initiatives of businesses such as Environmental Management Systems, codes of conduct, certification, and public reporting on environmental and social issues
- Set standards and targets for products and processes
- Establish programs of dialogue, partnership, and cooperation between individual businesses, key business sectors, communities, and government
- Education and training (especially for small and medium-sized enterprises) and
- Economic incentives such as full cost pricing, environmental taxes and fees, green tax reform, and the elimination of environmentally harmful subsidies should be used

To ensure sustainable consumer behavior:

- Sustainable public procurement
- Product labeling schemes
- Advertising and marketing
- Awareness campaigns
- More generally, the provision of information and training and
- Certain products need to be banned

Consumer and government interest in sustainable actions and policies is on the rise. For example, the number of countries covered by e-waste legislation and regulation increased from 61 in 2014 to 78 in 2019. The global e-waste recycling market is expected to grow from $8.4 billion in 2019 to an estimated $20.5 billion by 2027 (Tomsa Romonti-Maniu & Scridon, 2021). Individuals' consumption behavior plays a key role in the path to a sustainable future. Understanding what influences the decision to act in a sustainable way is therefore crucial. The influential factors that have been identified can be broadly divided into two categories: individual-related determinants and environmental determinants. The former includes socio-demographic, personal characteristics, and cognitive factors such as value orientation, non-cognitive factors (habits and emotions), and knowledge. Environmental determinants include product, service, or behavior-related factors (such as stereotypes of sustainable products), organizational activities (e.g., communication efforts), social impact, and structural conditions such as existing infrastructure (Wintschnig, 2021, p. 324).

9.2.5 Sustainable Consumption and Environmental Awareness

As a result of the unlimited and uncontrolled use of natural resources and production resources in the intensive production process that emerged with industrialization, it has been thought that measures should be taken to protect the environment with the increasing concern of managers about the environment since the 1970s. At the Conference on Environment and Man held by the United Nations in Stockholm in 1972, the Stockholm Declaration, the first international declaration on global environmental problems, was published by addressing the negative impact of industrialization and economic growth on the environment and the consumption of natural resources (Kayhan, 2013, p. 61). As mentioned in the report published by the United Nations in 1987, "sustainability"is a concept that first came to the agenda in the context of the environment. It is aimed to make the environment and nature sustainable and to realize the responsibilities toward nature (Tüyen, 2020, p. 91).

The environmental component of sustainability can be considered the main reason for adopting a voluntary simple lifestyle. Having an interest in protecting the environment can be reflected in behaviors to change individual actions, such as reducing one's consumption of goods and services only as much as one really needs (Seegebarth et al., 2016). Furthermore, such consumers engage in daily activities to protect the environment and reduce their carbon footprint to contribute less to global warming. For example, buying "green"or "eco-labeled" products is seen by consumers in various countries (UK, USA, France) as a way to reduce the impact of global warming (Roser-Renouf et al., 2016). If individuals place a high value on purchasing products made from recycled materials or packaging produced in an environmentally friendly way, they are more likely to purchase eco-labeled products or shop in stores that promote fair trade with the aim of contributing to sustainability (Maaya et al., 2018). If individuals are sensitive to issues related to recycling, eco-packaging, and environmentally friendly disposal of products, they are more likely to limit their consumption (food, energy, and clothing) to only what is needed. In this case, consumers will have contributed to reducing global warming and protecting the environment for future generations in the medium to long term (Whitmarsh et al., 2017).

Our world has undergone a significant transformation in the last 50 years, with global trade, massive consumption growth, and remarkable urbanization as a result of human population growth, and our way of life has changed dramatically. But this change has come at a great cost to nature and thus to the stability of our planet. The Ecological Footprint, designed by Mathis Wackernagel and William Rees at the University of British Columbia in 1990, is widely used by scientists, businesses, governments, individuals, and institutions to monitor ecological status, resource use, and advance sustainable development (https://www.footprintnetwork.org/). Since 1970, our Ecological Footprint has exceeded the Earth's capacity to renew itself, depleting the health of our planet and the future of humanity. Demand for natural resources is not evenly distributed across the globe. Since resources are no longer consumed where they are extracted, people's consumption of natural resources is shaped by their accessibility (WWF, 2012).

The unit of measurement for ecological footprint calculation is the "global hectare area (kha)". One hectare of biologically productive area is expressed in one kha. Thanks to this unit of measurement, all ecological footprint calculations can be made and explained in a consistent manner (Wilson & Anielski, 2005, p. 14). According to the best-case scenario, Fig. 9.1. shows a 27% decrease in the footprint of per capita consumption between 1961 and 2050. In the same period, biological capacity per capita decreased by 57%. The difference between biological capacity and footprint decreases considerably in the 2020s, and Turkey appears to be closing its ecological gap (WWF, 2012).

Individual Ecological Footprint varies across countries and provides information on countries' performance in using their resources and their risks and opportunities. The per capita Ecological Footprint is an indicator of total population and consumption status. Ecological Footprint levels are a result of different lifestyles and consumption patterns, such as the amount of food, goods, and services consumed by households, the amount of natural resources consumed, and the amount of carbon dioxide released into the atmosphere. The 2020 Living Planet Report's key message is: "Nature, our life support system, is deteriorating at an incredible rate." Severe forest fires and pandemics such as COVID-19, which threaten our health, prove that the health of humanity and our planet cannot be considered separately from each other (WWF, 2020).

The Living Planet Index tracks changes in the relative abundance of wild species populations over time. The global index is constructed by calculating an average trend for tens of thousands of land, freshwater, and marine vertebrate populations from around the world. The 2022 Global Living Planet Index (Fig. 9.2), which includes increasing and decreasing trends, shows an average decline of 69% in monitored populations between 1970 and 2018 (WWF, 2022, p. 32).

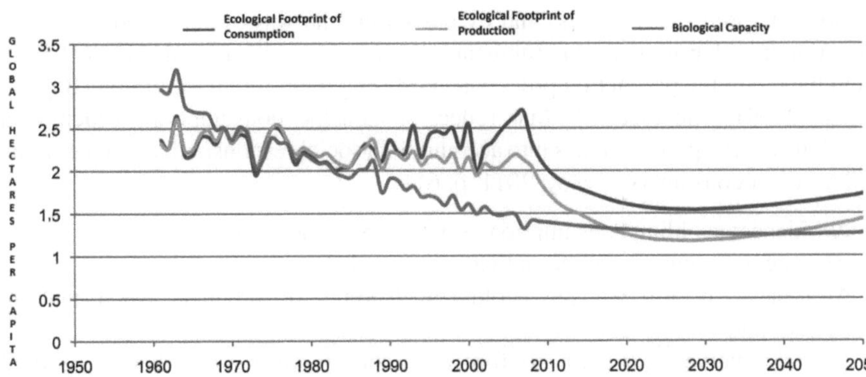

Fig. 9.1 Ecological footprint and biological capacity values per capita in the best-case scenario. Source: Turkey's Ecological Footprint Report (WWF, 2012)

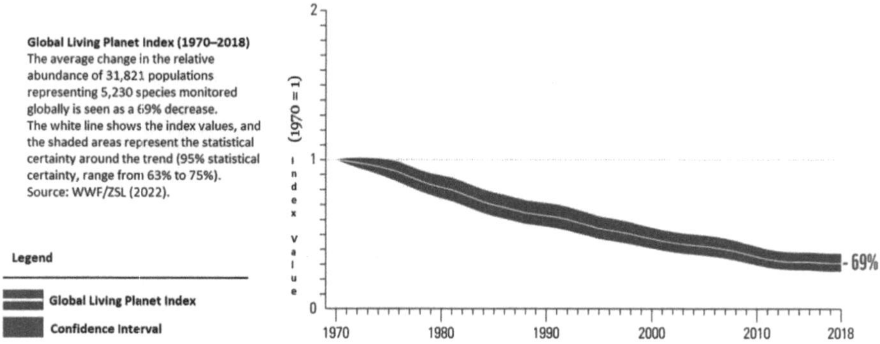

Fig. 9.2 Global living planet index. Source: WWF/ZSL, Living Planet Report (2022)

Considering the fact that our economies are not independent from nature, we can protect and develop biodiversity and contribute to our economic prosperity. Otherwise, catastrophic consequences for the environment, human health, and the economy are predicted. Protecting and improving our environment should be at the center of our understanding of economic success (WWF, 2020). Humanity's collective demand for resources, identified as the main driver of climate change, far exceeds the planet's natural capital. Therefore, finding ways to curb overconsumption is crucial for survival (Tobler et al., 2012, p. 197). Environmental concern and a general awareness of the consequences of harming nature are believed to be more pronounced among careful consumers, and there is a positive relationship between environmental concern and sustainable consumption (Helm & Subramaniam, 2019, p. 4). It is very important to understand the factors that determine consumer behavior when creating sustainable consumption awareness. It will be possible to support the sustainability of consumption by making natural resources ready for production through the production process and by preventing the wastes generated after consumption from harming the environment and production resources and by including them back into the production process and offering them for consumption (Bayazıt Hayta, 2009). The elements that reflect sustainable consumption, which is a decision-making process, takes into account both social responsibility and the wants and needs of consumers (Özsoy, 2011, p. 61):

- Avoidance: avoiding consumption or, where possible, consuming less
- Attitude; refusing consumption that is not needed
- Awareness; prioritizing the ecological features of products when making choices and
- The alternative is to turn to forms of consumption other than traditional consumption

The duties and responsibilities of a conscious consumer in protecting the environment can be evaluated as follows (Bener & Babaoğul, 2008, p. 8):

- Consumers can save money, energy, and time by identifying their priority needs and making planned purchases.
- Consumers should pay attention to the labeling and contents of the product during the purchase and prefer products that do not create harmful waste for nature and are made of recyclable materials.
- It is necessary to minimize household waste and make it recyclable. For example, discarded glass, plastic, and paper packaging can be collected separately from other waste and recycled.
- Since families play an important role in making their children conscious consumers, they should ensure that their children are aware of environmental protection.
- Consumers should cooperate in order to achieve effective results in protecting the environment. There are various associations, foundations, and organizations operating in this field in our country.

Research on sustainable consumption has evolved significantly from practice-oriented research focusing on a single issue to examining a range of interrelated issues (such as sustainability, consumption behavior, energy use, and environmental impact) (Liu et al., 2017, p. 414).

9.2.6 Conclusion

The study examined consumer behavior and found that raising awareness about individual consumption, social and environmental factors, and climate change are important issues in sustainability. Sustainable consumption has become a key concept for academics, institutions, and policy makers. Ongoing debates on climate change, economic growth, population growth, and related trends make sustainable consumption even more imperative for researchers (Lim, 2017, p. 69). From a sustainability perspective, there is a need for public policies and marketing and business systems that guide organizations and people to produce and consume within ecological limits and in a socially responsible manner (Prothero et al., 2011, p. 31). Sustainable consumption behavior involves an effort that has a positive impact on the lives of other people. Sustainable consumption is a complex and diverse field of scientific research that combines knowledge from different academic disciplines and creates assumptions for its translation into sustainable consumption policy and practice. The richness and diversity of scientific research in the field of sustainable consumption demonstrates the undeniable importance of the topic (Piligrimiene et al., 2019, p. 116).

Considering the increasing world population, it is envisaged that studies on alternatives and solutions for sustainable consumption should be increased and incentive practices should be implemented for both consumers and producers on sustainability. Organizations and consumers should be trained on the importance of environmental protection, climate change and sustainability, and awareness-raising activities should be carried out on what sustainability means for our planet and therefore for our future. It is thought that this study will provide ideas for future research on the role of consumers in ensuring sustainable consumption, as well as positioning other stakeholders, such as producers and governments, on sustainability.

References

Altıok N, Babaoğul M (2008) Sürdürülebilir tüketim ve tüketici eğitimi. 13th National marketing congress, Adana, pp 394–403

Aydoğan Boschele F (2020) Sürdürülebilirlik. Tüketim ve Medya İnsan ve İnsan 7(26):11–23. https://doi.org/10.29224/insanveinsan.757873

Bayazıt Hayta A (2009) Sürdürülebilir Tüketim Davranışının Kazanılmasında Tüketici Eğitiminin Rolü. Ahi Evran University Journal of Faculty of Education 10(3):143–151

Bener Ö, Babaoğul M (2008) Sürdürülebilir Tüketim Davranışı ve Çevre Bilinci Oluşturmada Bir Araç Olarak Tüketici Eğitimi. Hacettepe University Sociological Research E-Journal

Corral-Verdugo V (1997) Dual 'realities' of conservation behavior: self reports vs observations of re-use and recycling behavior. J Environ Psychol 17(2):135–145

Ergen A (2014) Maddi Değerler Gönüllü Sade Yaşam Biçimi, Çevre Bilgisi: Sürdürülebilir Tüketim Davranışı Açısından Bir Araştırma. (Unpublished Doctoral Thesis). Marmara University/ Institute of Social Sciences, Istanbul

Geiger SM, Otto S, Schrader U (2018) Mindfully green and healthy: An indirect path from mindfulness to ecological behavior. Front Psychol 8:2306

Gladwin TN, Kennelly JJ, Krause TS (1995) Shifting paradigms for sustainable development: implications for management theory and research. Acad Manag Rev 20(4):874–907

Gürbüz A Ayar İ (2018) Tüketicilerin Sürdürülebilir Satın Almaya İlişkin Motivasyonları Üzerine Bir Araştırma, Çankırı Örneği Int J Soc Res 11(60), Issn: 1307–9581

Hansen U, Schrader U (1997) A modern model of consumption for a sustainable society. J Consum Policy 20(4):443–468

Helm S, Subramaniam B (2019) Exploring socio-cognitive mindfulness in the context of sustainable consumption. Retailing and consumer sciences. University of Arizona, Tucson, AZ

IIED (1998) Consumption in a sustainable world: report of the workshop held in Kabelvag, Norway, Ministry of the Environment. Oslo and International Institute of Environment and Development, London

Karalar R, Kiracı H (2011) Çevresel Sorunlara Karşı Bir Çözüm Önerisi Olarak Sürdürülebilir Tüketim Düşüncesi. Dumlupınar University Journal of Social Sciences 30:63–76

Kayhan AK (2013) Birleşmiş Milletler Çevre Programı Üzerine Bir İnceleme. Public and Private International Law Review, Cilt 33(1):61–90

Kayıkçı P, Armağan K, Dal NE (2019) Sürdürülebilir Pazarlama: Kavramsal Bir Çalışma. Burdur Mehmet Akif Ersoy University, Bucak Faculty of Business Journal 2(1):77–93

Keleş R, Hamamcı C (1993) Çevrebilim, 50. İmge Kitapevi, Ankara

Kong N, Salzman O, Steger U, Ionescu-Somers A (2002) Moving business/industry towards sustainable consumption: the role of NGOs, s.109. Eur Manag J 20(2)

Li S (2003) Recycling behavior under chinas social and economic transition, the case of Metropolitan Wuhan. Environ Behav 35(6):784–801

Lim WM (2017) Inside the sustainable consumption theoretical toolbox: critical concepts for sustainability, consumption, and marketing. Malezya, Journal of Business Research, Cilt 78:69–80

Liu Y, Qu Y, Lei Z, Jia H (2017) Understanding the evolution of sustainable consumption research. Sustain Dev 25:414–430

Maaya L, Meulders M, Surmont N, Vandebroek M (2018) Effect of Environmental and Altruistic Attitudes unwillingness-to-Pay for Organic and Fair Trade Coffee in Flanders. Sustain J Rec 10:4496

Marın MC, Yıldırım U (2004) Çevre Sorunlarına Çağdaş Yaklaşımlar, 253. Beta, İstanbul

McLaren SJ (2007) Defining a role for sustainable consumption initiatives in New Zealand. 2nd International conference on sustainability engineering and science, Auckland, NZ

Niva M, Makela J, Kahma N, Kjærnes U (2014) Eating sustainably? Practices and background factors of ecological food consumption in four Nordic Countries. J Consum Polic 37:465–484

OECD (2002) Towards sustainable household consumption? Trends and policies in OECD countries. Access address: https://read.oecd-ilibrary.org/environment/towards-sustainable-household-consumption_9789264175068-en#page12

Özgül E (2010) Tüketicilerin Değer Yapıları, Gönüllü Sade Yaşam Tarzı Ve Sürdürülebilir Tüketim Üzerindeki Etkileri. Hacettepe University Journal of Faculty of Economics and Administrative Sciences 28(2):117–150

Özsoy T (2011) Tüketimin Sürdürülebilirliği: Ürün Ömrüne Yönelik Tüketici Tutum ve Davranışları Üzerine Bir Araştırma, 61, Çukurova University/Institute of Social Sciences (Unpublished Doctoral Thesis), Adana

Piligrimiene Z, Banyte J, Virvilaite R (2019) Work-life balance and sustainable consumption: is there a Relation. Transformations in Business and Economics 18(3(48)):116–127

Prothero A, Dobscha S, Freund J, Kilbourne WE, Luchs MG, Ozanne LK, Thogersen J (2011) Sustainable consumption: opportunities for consumer research and public policy. J Public Policy Mark 30(1):31–38

Roser-Renouf C, Atkinson L, Maibach E, Leiserowitz A (2016) Climate and sustainabilityl The consumer as climate activist. Int J Commun 10:24

Seegebarth B, Peyer M, Balderjahn I, Wiedmann KP (2016) The sustainability roots of anticonsumption lifestyles and initial insights regarding their effects on consumers' well-being. J Consum Aff 50:68–99

Şener A, Hazer O (2008) Values and sustainable consumption behavior of women: a Turkish sample. Sustain Dev 16:291–300

Seyfang G (2006) Ecological citizenship and sustainable consumption: examining local organic food network. J Rural Stud No. 22

Taluğ C, Birden B (2014) Sürdürülebilir Üretim ve Tüketim. Türkiye Biyoetik Dergisi 1(2), 78–91. Compilation

Tobler C, Visschers VH, Siegrist M (2012) Addressing climate change: determinants of consumers' willingness to act and to support policy measures. J Environ Psychol 32(3):197–207

Tomsa MM, Romonti-Maniu AI, Scridon MA (2021) Is sustainable consumption translated into ethical consumer behavior? Sustain For 13(6):3466

Tunçluer H (2010) Sürdürülebilirlik: Kalkınma, Pazarlama, Mutlak Organik Tekstil. 1. Baskı. Tunçluer Yayınları, İstanbul

Tüyen Z (2020) İşletmelerde Sürdürülebilirlik Kavramı ve Sürdürülebilirliği Etkileyen Etmenler. Istanbul Commerce University Journal of Social Sciences, Issue 37:91–117

United Nations Environment Programme (2001) UNEP annual report 2001. https://wedocs.unep.org/20.500.11822/8168

Vermeir I, Verbeke W (2006) Sustainable food consumption: Exploring the consumer "attitude–behavioral intention" gap. J Agric Environ Ethics 19(2):169–194

Vermeir I, Verbeke W (2008) Sustainable food consumption among young adults in belgium: theory of planned behaviour and the role of confidence and values. Ecol Econ 64(3):542–553

Visvanathan C Kumar S (2007) 3R initiatives in Asia, S.11, In International conference on cleaner technologies and environmental management (ICCTEM)

Whitmarsh L, Capstick S, Nash N (2017) Who is reducing their material consumption and why? A cross-cultural analysis of dematerialization behaviours. Philos Trans R Soc A Math Phys Eng Sci 375:20160376

Wilson J, Anielski M (2005) Ecological footprints of Canadian municipalities and regions, 14. Anielski Management Inc, Edmonton

Wintschnig, B.A. (2021). The Attitude-Behavior Gap – Drivers and Barriers of Sustainable Consumption. Junior Management Science, 6(2), 324–346

WSSD (2002) Report of the world summit on sustainable development. Johannesburg, South Africa. ISBN 92-1-104521-5

WWF (2012) World Wildlife Fund, Living planet report 2012: Biodiversity, biocapacity and better choices, Gland, 6

WWF (2020) World Wildlife Fund, Living planet report 2020: biodiversity, biocapacity and better choices. Gland

WWF (2022) World Wildlife Fund, *Living Planet Report 2022*: Biodiversity, biocapacity and better choices. Gland, p 32

Young W, Hwang K, McDonald S Oates CJ (2010) Sustainable consumption: Green consumer behaviour when purchasing products. Sustain Dev, 18(1), 18–31. https://www.footprintnetwork.org/our-work/ecological-footprint/#worldfootprint (Access date: 10.02.2022)

Aysun Duruş, born in 1987 in Kırşehir, Turkey, completed her high school education at Kırıkkale Anadolu High School. She earned a Bachelor's degree in Business Administration from the Faculty of Economics and Administrative Sciences at Kırıkkale University in 2010. After graduation, she worked as a Customer Relations Officer at Kastamonu Ziraat Bankası A.Ş., gaining significant experience in sales, marketing, and customer relations. After 2 years of banking experience, she began her career as a civil servant, working at Aksaray University, Konya Selçuk University, and Ankara Gazi University. In 2023, she completed her Master's degree with a thesis at the Department of Business Administration, Faculty of Economics and Administrative Sciences, Tekirdağ Namık Kemal University. Her thesis, titled "Analysis of Sustainable Consumption Studies from the Consumer Perspective with Text Mining," was supervised by Associate Professor Dr. Şeniz Özhan and Asist. Prof. Erkan Özhan. She is currently working as a civil servant at the same university.

Şeniz Özhan was born in Istanbul in 1979. She completed her undergraduate education in the Department of Business Administration at the Faculty of Economics and Administrative Sciences at Trakya University in 2003, her master's degree in the Department of Business Administration at the Institute of Social Sciences of the same university in 2007, and her doctorate in 2015. She worked at Tekirdağ Namık Kemal University's Çorlu Vocational School between 2005 and 2023. Since 2023, she has been working as an Associate Professor in the Department of Business Administration at the Faculty of Economics and Administrative Sciences of Tekirdağ Namık Kemal University, specializing in the Department of Production Management and Marketing. Her research focuses on consumer behavior, service marketing, marketing research, sustainable consumption, and contemporary marketing approaches.

Part V
New Approach in Banking and Economy

Chapter 10
Analysis of the Labor Market in Turkey with a Focus on Inclusive Growth

Emel Yıldız and Nural Yıldız

Abstract Inclusive growth is a type of economic expansion where prosperity is distributed equitably across society, ensuring equal access to opportunities and benefits for all. The inclusivity of growth depends on factors such as the education and employment of the population, their participation in production, and the quality and quantity of value added during the production process. Low labor force participation, high unemployment, and inequalities in income distribution and labor markets during this process reveal the presence of injustice in sharing the benefits of economic growth. Inequalities in the economy not only affect individual gains but also result in losses to societal well-being. Inclusive growth, on the other hand, serves as a fundamental basis for the sustainability of development.

The aim of this study is to analyze the labor market in Turkey, emphasizing employment and equality as essential elements for the sustainability of societal well-being. Based on the analysis conducted using data from the Turkish Statistical Institute (TÜİK), the study suggests that addressing existing inequalities, starting with disadvantaged groups, and increasing labor force participation and employment rates are crucial for promoting inclusive growth and enhancing societal well-being.

Keywords Inclusive growth · Employment · Unemployment · Income distribution

E. Yıldız (✉)
Faculty of Economics and Business Administration, Division of Cooperative Management, Edirne, Turkiye
e-mail: emelyildiz@trakya.edu.tr

N. Yıldız
Faculty of Economics and Business Administration, Division of Economics, Edirne, Turkiye
e-mail: ynural@trakya.edu.tr

K. T. Çalıyurt (ed.), *New Approaches to CSR, Sustainability and Accountability, Volume VI*, Accounting, Finance, Sustainability, Governance & Fraud: Theory and Application, https://doi.org/10.1007/978-981-96-5937-1_10

10.1 Introduction

Economic activities that aim not only to meet individuals' needs but also to enhance their well-being primarily focus on production, consumption, trade, and distribution. Ensuring fair access for individuals to all these areas not only ensures the effectiveness of this process but also makes the increase in well-being sustainable. From the point where the First Industrial Revolution is considered a milestone for economic growth, spanning from that moment to the twenty-first century, there have been significant increases in material indicators such as production and income. While the share taken from the distribution of this material development creates wealth, its unjust distribution has led to or deepened poverty with economic, social, political, and environmental dimensions.

The First Industrial Revolution, accepted as the turning point for economic growth, raised concerns about the unequal distribution of wealth and the inclusiveness problem of growth along with developments in production and income. The global average of GDP per capita, which was 666 dollars in 1820, increased to 7614 dollars in 2008. In the UK (Great Britain), the GDP per capita rose from 1706 dollars in 1820 to 23,742 dollars in 2008, while sub-Saharan African countries could only reach 1387 dollars from 415 dollars during the same period (OECD, 2006). The income gap between the lowest and highest income groups, which was threefold until the First Industrial Revolution, increased to one hundred fifty times with the Industrial Revolution and five hundred times with the technological revolution (Yaz, 2021:116). In terms of income distribution, inequalities, presented as an economic problem, also acquire a social dimension by undermining social peace and trust. At this stage, inclusive growth represents an approach aimed at eliminating inequalities in income generation, access to opportunities, and welfare distribution.

In Turkey, alongside labor force participation and income distribution inequalities, these problems also have a separate dimension concerning gender distinctions. According to the World Inequality Report (2022), the top 10% of the population in Turkey, which earns the most, has income 23 times higher than the bottom 50% of the population. In other words, the richest 10% of the population, constituting 10%, takes 54.5% of the total income, while the bottom 50% of the population takes only 12% of the total income. In 2022, according to the income distribution statistics of the Turkish Statistical Institute (TÜİK), the share of men in total labor income is 43.25% higher than the share of women (TÜİK, 2023). Income inequality, in addition to its societal dimension, extends to sub-dimensions such as working poverty and women's poverty.

The OECD Growth and Development Commission's 2008 Growth Report emphasized the necessity of reducing poverty alongside high growth and the sustainability of growth. Equal opportunity, equality, and employment are fundamental components of inclusive growth, while unequal opportunities are identified as a "toxic" element damaging the growth process.

Labor markets reflect many important cause-and-effect relationships in terms of education, labor force participation, employment rate, and wage policies. Properly constructing these relationships maximizes individual and social benefits while ensuring the inclusiveness of development. In this study, the labor market in Turkey has been considered in a general sense using TÜİK (Turkish Statistical Institution) data from the perspective of inclusive development. For the inclusiveness and quality of growth to increase, it is important to implement employment and wage policies that prioritize labor productivity, supporting the participation of women and young people in the labor force first.

10.2 Economic Growth to Inclusive Growth

In 1690, English scientist William Petty, in his book *Political Arithmetic*, conducted calculations to measure England's economic well-being using data such as population, land value, personal income, and personal expenditure. This approach, laying the foundation for the science of economics, gained clarity in the 1930s with the development of the Gross National Product (GNP) concept by Australian economist Colin Clark (Abbot et al., 2012:36–37). The increase in GNP during a specific period has been expressed as economic growth.

Economic growth denotes the long-term increase in production per capita in a country (Kuznets, 1973:24). Economic growth occurs through an increase in the quantity of production factors or the efficiency of production factors. While qualitative improvements may not necessarily demonstrate individuals' welfare, the idea that economic growth is necessary is generally accepted for the enhancement of well-being.

Despite increases in GNP worldwide, issues such as poverty and unemployment have also risen, shifting approaches from those measuring economic growth through quantitative increases in production and income to approaches sensitive to issues such as income distribution justice, poverty, unemployment, environmental problems, democracy, and cultural changes. In other words, the focus has shifted to the inequalities in the distribution of prosperity achieved through economic growth among countries and individuals. With this perspective, the quality of economic growth, the number of people it includes or excludes, and similar issues have started to be addressed through the 'inclusive growth' approach.

Inclusive growth allows individuals to actively contribute to the quality of life and societal progress. The 'inclusive growth'approach emphasizes the quality of growth by focusing on inequalities in the distribution of prosperity (OECD, 2012:280).

OECD (2019:135) has identified two fundamental dimensions for inclusive growth:

- Adoption of a regional approach; a development approach that facilitates spreading from developed to less developed areas, especially for regions in the industrial transition process.
- Facilitating industrial transition with appropriate labor market and skill development policies for vulnerable groups and those most affected by industrial change.

Shifting public infrastructure spending to underdeveloped regions, addressing education, labor force participation, and employment policies starting with disadvantaged or marginalized groups, is a prerequisite for inclusiveness to rapidly and sustainably reducing poverty. In addition to its economic importance, the inclusiveness of growth is a necessary condition for the sustainability of development, social peace, and harmony.

10.3 The Relationship Between Employment and Economic Growth

The crucial point for countries is not only the quantity of factors such as labor, capital, natural resources, entrepreneurship, and technological levels but also the high level of their employment and the efficiency of the production process. The effective and high-level employment of resources serves as a driving force for countries' development, whereas inadequately or improperly utilized resources can lead to a decrease in societal welfare. Besides employment, the enhancement of productivity through education is also a key concern in development, as labor, the fundamental input in the production process, contributes significantly. Developed countries are characterized by high levels of education and employment at the societal level, while developing countries exhibit lower levels of education and employment, with women lagging behind compared to men.

Economist Alfred Amonn stated that the country's economy would undergo changes in two directions over time (Cited in: Acar, 2008:9):

1. The body of an economy growing with its population: Population growth leading to an increase in workforce or production factors. This process represents quantitative growth and is referred to as economic growth.
2. The structure and framework of an economy changing: Changes in the shares of agriculture, industry, and service sectors within the GDP, the sectoral distribution of the workforce, and the country's production infrastructure. This process signifies qualitative development and structural change, i.e., development, which is a more comprehensive concept encompassing growth.

The UN Human Development Report (1996) expressed caution against "Bad Growth," which fails to create sufficient employment opportunities, distributes increasing income unfairly, does not improve democracy, rights, and freedoms, leads to the loss of cultural values, traditions, and ecological imbalance. In contrast, "Good Growth" is defined as growth that promotes employment, shares prosperity fairly, enhances democratic rights and freedoms, preserves cultural values, fosters social harmony, cooperation, and solidarity, and safeguards the environment for future generations (UNDP, 1996).

In economies, the term 'jobless growth' is used when unemployment does not decrease proportionately despite economic growth. The inability to generate sufficient employment or unemployment leads to:

- Negative Deviation from Potential GDP
- Increase in Public Expenditures
- Idle Workforce
- Rise in Idle Capacity and Socio-Economic Effects

The relationship between unemployment and growth is extensively studied by many economists. For instance, Arthur Okun examined the relationship between growth and unemployment in the American economy from 1948 to 1960. He determined, in what is now known as 'Okun's Law,' that during periods of high real growth, the unemployment rate decreases, and conversely, during periods of decreasing real growth, unemployment increases (Okun, 1962:98–104).

10.4 Inclusive Growth and the Significance of Employment: Realizing Potential and Employment Data in Turkey

For the inclusiveness of growth, the active participation of the population in the workforce, the quality and quantity of employment, and the value it adds to the production process are crucial. Countries' growth policies should be shaped based on productivity, innovation, and equality, with a focus on achieving high levels of employment by carefully analyzing their own conditions.

In Turkey, out of the 65,494,000 individuals aged 15 and above who are eligible to work (TÜİK, 2023), only 35,044,000 are actively participating in the workforce, as seen in Table 10.1. The overall labor force participation rate of 53.6% decreases to 36.1% for women. Besides lower participation rates for women, the unemployment rate is higher for women at 13%, compared to 7.7% for men. Similarly, in youth unemployment (15–24 age group), the rate for women is higher at 22.3% compared to men.

In Table 10.2, the ratio of employees covered by social security to the workforce in Turkey is 74.08%, indicating the existence of informal employment. Informal employment, not included in GNP calculations, remains outside social protection and similar rights.

Societal encouragement of work and productivity, and the disapproval of idleness, are important. However, it is equally crucial for individuals participating in the workforce to find satisfaction in their work and wages. Job satisfaction is not only derived from the joy of being productive and effective in one's work but is also motivated by satisfaction with wages.

Table 10.1 Labor Force Indicators in Turkey (May, 2023) (Ages 15 and above, Seasonally Adjusted)

Thousand people	Total	Female	Male
Labor Force	35.044	11.909	23.135
Employment	31.716	10.359	21.357
Unemployed	3.328	1.549	1.778
Not in the Labor Force	30.338	21.096	9.212
Percentage	**Total**	**Female**	**Male**
Labor Force Participation Rate	53.6	36.1	71.5
Employment Rate	48.5	31.4	66
Unemployment Rate	9.5	13.0	7.7
Youth Unemployment Rate, Ages 15–24	17	22.3	14.3

Source: TURKSTAT, 2023 labor force statistic

Table 10.2 Ratio of employees registered within the scope of social security to the labor force in Turkey (2022)

Labor force	34.334.000
Actively Insured (Registered Employment)	25.435.806
Ratio of Registered Employees to Labor Force (%)	74.08

Source: SGK Annual Statistics www.sgk.gov.tr, TURKSTAT Labor Force Statistics www.tuik.gov.tr

Challenges such as employee poverty, unprotected employment, gender-based wage disparities, and sexist societal pressures are reasons that deter individuals from participating in the labor market. Special consideration and positive discrimination for vulnerable groups, particularly women, regarding the quality and security of employment are essential to reduce poverty and enhance the inclusiveness of growth. Additionally, employment should be linked with productivity and innovation.

Ragnar Nurkse (1953) expressed the perpetuation of poverty and unemployment as an ongoing condition without intervention, defining it as a Vicious Circle Theory. The poverty vicious circle suggests that poverty leads to low income and, consequently, low savings, investments, productivity, and employment. Ultimately, this causality will again lead back to poverty, initiating economic developments from the starting point of poverty. Thus, eliminating unemployment and addressing employee poverty are crucial findings (Fig. 10.1).

In Turkey, the implementation of a minimum wage that forms the basis for all incomes can be considered within the context of the poverty vicious cycle. The new minimum wage for the Second Six Months of 2023 is 11,402 TL (https://www.csgb.gov.tr/haberler/2023), and as of November of the same year, the poverty line in Turkey was determined as 45,000 TL, and the hunger line was determined as 13,000 TL (BİSAM, November 2023).

The failure to establish a higher income level as part of income policy raises the alternative cost of women's participation in working life, leading to their lower participation. A low minimum wage leaves women insufficient to cover expenses such as transportation, childcare, and clothing, resulting in the exclusion of women

Fig. 10.1 Vicious circle of poverty. Source: Nurkse (1953) Problems of Capital Formation in Underdeveloped Countries. Oxford: Oxford University Press

from the labor market. A rising wage parallel to labor productivity will allow for increased incomes for workers and households, higher savings potential, and better allocation of resources for the education and development of their children.

Table 10.3 illustrates significant differences in annual average basic income based on education levels and gender, both between different education levels and between women and men with the same education level.

Examining basic income by sector in Table 10.4, the highest annual average income is observed in the service sector at 66,072 TL, followed by the industrial sector at 62,200 TL, the construction sector at 59,926 TL, and the lowest annual average income is in the agricultural sector at 46,106 TL (TÜİK, 2022).

In Table 10.4, employers receive the highest income based on basic income, while daily workers receive the lowest, with women consistently receiving less than men across all income groups. It is evident that in the Turkish labor market, there is a need for more support for women in terms of inclusivity.

10.5 Women's Employment and Inclusivity in Turkey

During her tenure as a Member of Parliament in New Zealand, Marilyn Waring brought attention in 1988 to the underrepresentation of women's work in Gross Domestic Product (GDP) calculations in her work titled 'If Women Counted'

Table 10.3 Annual average main job income by education level (TL) 2022

Education status	Total	Female	Male
İlliterate	25,406	21,181	30,799
Non-graduate	34,755	23,141	39,994
Pre-high school education	46,916	30,233	50,953
High school and equivalent school	62,835	40,276	69,085
Higher education	87,451	68,026	99,836

Source: TURKSTAT; Income Distribution Statistics, 2022

Table 10.4 Annual average main business income (TL), 2022

Situation at work	Total	Female	Male
Paid, salaries	56,899	48,867	60,541
Daily wage	25,632	14,784	28,651
Employer	220,482	180,938	225,970
Own Account	58,697	36,543	62,556

Source: TURKSTAT; Income Distribution Statistics, 2022

(Abbot et al., 2012:310–311). At this stage, the contribution of women to activities such as childcare and domestic services, which are not purchased from the market but contribute to the family economy, is often overlooked. Individual rights violations, such as the loss of income and retirement rights when women do not actively participate in the workforce, are also ignored. On the other hand, women who actively participate in the workforce continue to engage in unpaid work, known as the 'second shift,' once they return home. The lack of market value for the responsibilities of women, such as childcare and housekeeping, renders women's domestic labor invisible and marginalizes them.

The participation of women in the production process, the employment of female labor, the analysis of employment, and the development of labor markets and the economy are essential processes. There are numerous economic, social, and political factors that determine employment. It is well-known that gender equality nurtures economic growth, and economic growth, in turn, fosters gender equality. Education and employment are dynamic factors that nourish gender equality, contributing to societal gains in the form of workforce productivity, child health, family well-being, and positive outcomes for future generations (Kabe, 2013:1). The increase in women's education and employment has significant effects supporting social transformation. However, growth that does not rely on gender equality cannot be sustainable and inclusive.

According to data from the Turkish Statistical Institute (TÜİK), there are differences in women's labor force participation in Turkey based on factors such as educational level, region, and cultural structure.

- The percentage of the population aged 25 and above with university degrees (associate, bachelor, master's, and doctoral) was 23.0% in 2021. The percentage for women increased from 7.6% in 2008 to 20.9% in 2021, while for men, it increased from 12.1% in 2008 to 25.1% in 2021.

- The labor force participation rate increases as the education level of women rises. While the labor force participation rate for illiterate women is 12.8%, it is 67.6% for women with higher education.
- The highest female employment rate, 36.8%, is observed in the TR90 region (Trabzon, Ordu, Giresun, Rize, Artvin, Gümüşhane).
- The lowest female employment rate is 14.5% in the TRC3 region (Mardin, Batman, Şırnak, Siirt), followed by TRC2 (Şanlıurfa, Diyarbakır) with 15.9%, and TRB2 (Van, Muş, Bitlis, Hakkari) with 19.6%.
- There are regional disparities in female employment.
- According to family structure research in 2021, 82.6% of individuals believe that women's work and contribution to social life are valuable. This perception is 85.6% among women and 79.5% among men.
- In 2021, the employment rate for women aged 25–49 with children under the age of 3 is 26.1%, while for men, it is 89.1%.

As seen in Table 10.5, the issue of women's employment is related not only to economic aspects but also to ingrained mentalities such as traditions and gender roles. In this regard, regions in Turkey with a more traditional structure, particularly in the eastern regions, exhibit lower female employment rates compared to the northern and western regions.

When household chores are examined based on gender, it is observed that women generally undertake household tasks in Turkey. Women are responsible for tasks such as childcare (94.4%), laundry and dishwashing (85.6%, even with the use of machines), cooking (85.4%), daily house cleaning, while men are more involved in tasks like paying monthly bills (74.1%), minor maintenance, repair, and fixing (65.2%) (TÜİK, 2022).

The inclusivity of growth requires addressing inequalities by focusing on disadvantaged groups, alleviating poverty, and increasing prosperity, all while concentrating on fair distribution. Policies that actively and qualitatively involve socially

Table 10.5 Perceptions about women's employment by gender (%)

	Female	Male
It is valuable for women to work and contribute to social life	85.6	79.5
Working women also show the necessary attention to their children	73.5	66.0
For a healthy family life, a woman should have a regular job	50.4	35.7
The main duty of women is child care and housework	31.5	40.2
Women should not work unless it is necessary for family income	44.0	56.0
It is against our traditions for women to work	11.2	16.8

Source: Turkey Family Structure Survey, 2021

vulnerable groups, such as women, the elderly, disabled individuals, and children, in social and economic life will ultimately enhance societal gains.

The main reasons behind women's unequal access to welfare include low rates of labor force participation, exposure to gender-based inequalities in working conditions, the exclusivity of traditional and cultural values, inadequate knowledge about how to integrate into the workforce, and issues related to access to finance.

The widespread availability of accessible and quality care services, the development of options such as paid leave and childcare leave, and the implementation of a cultural change approach supporting awareness and consciousness regarding equal participation of men in childcare and household chores will significantly contribute to solutions.

10.6 Conclusion and Recommendations

In the process of inclusive growth, where every segment of society can participate equally, the prosperity gained through growth is fairly shared and issues such as unemployment, poverty, and income inequality are minimized. The quality of growth, in this context, is achieved by creating employment, gender equality, equal opportunities, and income justice.

Examining the labor market in Turkey:

- Labor force participation is generally low, especially among women.
- There is a significantly high untapped labor force, resulting in individual and national growth and welfare losses.
- The employment rate is low, and the unemployment rate is high, especially among the youth and women, indicating that the economy is not creating sufficient employment, which is discouraging for young people and their families.
- There are significant differences in income among workers based on employment status, education, and gender, and the minimum wage is below the poverty line.
- Societal values and an understanding that assigns tasks such as childcare, health care, and elderly care to women contribute to making women's labor at home invisible. This issue is not only an economic problem but also a social and cultural problem that requires a solution.

For the inclusivity of development, it is crucial to implement programs that increase labor market participation, employment rates, gender equality, and productivity. If labor markets cannot be improved in terms of justice, efficiency, and increased employment, the vicious cycle in Turkey will pose a socio-economic threat to the sustainability of growth. Therefore:

- Integrating the vulnerable population into the labor market,
- Supporting disadvantaged groups (long-term unemployed, women, and youth) through career guidance and retraining programs, especially for young entrants and the unemployed,

- Creating inclusive employment opportunities associated with productivity, innovation, and competitiveness,
- Supporting investments to eliminate regional development disparities by considering sectoral potentials, and
- Planning and implementing industrial transformation to break the cycle of poverty through collaboration between universities, industry, local governments, and NGOs, for instance, achieving agricultural-based industrialization to eliminate urban-rural divides.

The discovery and effective utilization of potential are crucial steps for the well-being of individuals and society. Producing individually and nationally, redistributing and producing a better one is the basis of modernization and inclusive growth.

References

Abbot G et al (2012) The economics book. Kindersley, London

Acar Y (2008) İktisadi Büyüme ve Büyüme Modelleri, 5. Dora Yayıncılık, Baskı, Bursa

BİSAM - DİSK Birleşik Metal-İş Sınıf Araştırmaları Merkezi, https://www.dunya.com/ekonomik-veriler/yoksulluk-siniri (Accessed on November 18th, 2023).

Kabe N (2013) Paid work, women's empowerment and inclusive growth transforming the structures of constraint, UN.

Kuznets S (1973) Modern economic growth: findings and reflections. Am Econ Rev 63/3

Nurkse R (1953) Problems of capital formation in underdeveloped countries. Oxford University Press, Oxford

OECD (2006) Angus Maddison, The World Economy. http://www.ggdc.net/maddison (Accessed on November 10th, 2023)

OECD (2012) OECD Economic Surveys: Turkey 2012, OECD Publishing, Paris. https://doi.org/10.1787/eco_surveys-tur-2012-en (Accessed on March 18th, 2024)

OECD (2019) Regıons in industrial transition: policies for people and places. https://www.oecd-ilibrary.org, p.135 (Erişim: 10.10.2024)

Okun AM (1962) Potential GNP: its measurement and significance. American, statistical association, proceedings of the business and economics statistics section, pp 98–104

TÜİK, İstatistiklerle Kadın (2022). https://data.tuik.gov.tr/Bulten/Index?p=Istatistiklerle-Kadin-2022-49668 (Accessed on November 17th, 2024)

TÜİK, Gelir Dağılımı İstatistikleri, 04 Mayıs 2023 Haber Bülteni, 2022 https://data.tuik.gov.tr/Bulten/Index?p=Gelir-Dagilimi-Istatistikleri-2022-49745 (Accessed on March 20th, 2024)

UNDP (1996) Human Development Report. https://www.tr.undp.org/content/turkey/tr/home/library/national-hdrs/nhdr- 1996.html (Accessed on 8.05.2024)

Yaz, D. A. (2021) Antik Çağdan Geleceğe Para, 3. Baskı, İstanbul: Timaş Yayınları

Emel Yildiz born in 1971 in Ankara, Türkiye. In 1988, she graduated from Edirne High School and in 1993 she graduated from Marmara University, Faculty of Economics and Administrative Sciences, Department of Economics. In 1997, she completed her master's degree in Economics Program at Trakya University Institute of Social Sciences with the thesis "An Analysis on Employment Problems of Industrial Enterprises in Kırklareli." In 2006, she received her Ph.D. degree from Trakya University Institute of Science and Technology, Agricultural Economics Program with the thesis "Econometric Analysis of the Effects of Monetary and Fiscal Policies on Agricultural Production." In 2007, she received the title of assistant professor and in 2019 he received the title of associate professor in the field of Macroeconomics. Her research interests are human capital, sustainability, economic growth, and development. She has a special interest in cooperatives and the place of women in socio-economic life.

Nural Yildiz born in 1969, Uzunköprü, Türkiye. In 1986, he graduated from Edirne High School and in 1992 he graduated from Dokuz Eylül University, Faculty of Economics and Administrative Sciences, Department of Economics. In 1997, he completed her master's degree in Economics Program at Trakya University Institute of Social Sciences with the thesis "Economic Analysis of Water Resources in Thrace." In 2005, he received his Ph.D. degree from Trakya University Institute of Science and Technology, Agricultural Economics Program with the thesis "Economic analysis of employment, inflation, growth developments in the Turkish economy and the effects of sectors on these macro variables." In 2006, he received the title of assistant professor and in 2019 he received the title of associate professor in the field of Macroeconomics. His research interests are Turkish Economy, employment, and agricultural economics.

Chapter 11
Project-Planning Mode and Its Relation to Team Characteristics and Project Outcomes

Said Elbanna and Mustafa Çolak

Abstract Based on self-efficacy, team potency, group dynamics, and normative decision theories, we argue in this study that certain project team characteristics, namely, team management skills and team stability, affect the project-planning mode and that planning mode is related to the speed of project completion and project disturbance. We summarize our arguments under five hypotheses, one of which postulates a mediation effect between the two project outcomes. The analysis of 410 projects from the UAE reveals that adopting a more deliberate mode in planning projects is more likely to succeed, and consequently, it demonstrates the importance of normative decision theory as a theoretical lens in the context of planning projects. Our findings offer some implications for both academia, which we discussed in relation to theoretical lenses and contextual variables employed in this study, and practice, which we discussed in relation to team stability and team building and development.

Keywords Project-planning mode · Emergent/deliberate planning · Project disturbance · Speed of project completion · Team stability and skills

Said Elbanna and Mustafa Çolak contributed equally with all other contributors.

S. Elbanna
Department of Management and Marketing, College of Business and Economics, Qatar University, Doha, Qatar
e-mail: selbanna@qu.edu.qa

M. Çolak (✉)
Social Sciences University of Ankara, Ankara, Türkiye
e-mail: mustafa.colak@asbu.edu.tr

K. T. Çalıyurt (ed.), *New Approaches to CSR, Sustainability and Accountability, Volume VI*, Accounting, Finance, Sustainability, Governance & Fraud: Theory and Application, https://doi.org/10.1007/978-981-96-5937-1_11

11.1 Introduction

Projects can be planned in various ways ranging from emergent mode to planned (or deliberate) mode (Sirén & Kohtamäki, 2016; Titus et al., 2010). This classification of planning mode as "emergent" vs. "deliberate" has generated significant amount of scholarly debate in strategy research. In the context of project management which of the two modes of project planning is most beneficial for project outcomes remains debatable. Although relevant published findings are inconsistent (Guo et al., 2020; Serrador & Turner, 2015b), there is a consensus today that every project, except for those that are trivial, such as small-scale and short-term projects that involve few tasks and a small budget, requires some degree of planning and that planning constitutes a key element of modern project management (Dvir et al., 2003).

Despite the consensus concerning the importance and benefits of planning, in general, and project planning, in particular (Dvir & Lechler, 2004; Dvir et al., 2003; Elbanna & Elsharnouby, 2018; O'Shannassy, 2017), most studies on project planning are largely based on conjecture, with few empirical data produced concerning its effect on project outcomes (Hoegl & Parboteeah, 2006). Furthermore, the available empirical research on strategic planning, in general, and project planning, in particular, are inconsistent (Serrador & Turner, 2015b), and there is ongoing debate over the effect of planning on both organizational- and project-level outcomes.

The present study aimed to settle the project planning debate by testing the hypotheses that the choice of project-planning mode is affected by various team-level factors (namely, project management skills and team stability) and that planning mode affects project outcomes (namely, speed of project completion and project disturbance). So, in contrast to early studies that have exclusively focused on planning-outcome relationship (e.g., Dvir et al., 2003; Serrador & Turner, 2015b), the present study explicitly focuses on project-planning mode (i.e., planning process) and its relationship to some team-level determinants and less examined project-level outcomes, namely, project disturbance.

Similarly, our proposed study should be a valuable effort because, compared to strategic management literature, where the antecedents and outcomes of strategy process are empirically well-examined (see for example, Shepherd & Rudd, 2014; Wolf & Floyd, 2017), project management literature offers, to the best of our knowledge, less research on this issue. Even though some recent studies (e.g., Dayan et al., 2012; Elbanna, 2015) developed comprehensive models that integrate various contextual variables on the project, firm, and environmental levels, and examined how these variables are related to intuition and political behavior, they have at least two shortcomings, as compared to the present study. First, these earlier studies considered intuitive and political decision-making as proxies for emergent planning and did not explicitly focus on the planning mode (emergent vs. deliberate). Second, they were concerned primarily about how project-, firm-, and environmental-level variables affect intuitive and/or political processes (i.e., emergent mode of planning), and project outcomes (e.g., speed of completion, project success, etc.), ignoring the effect of team-level antecedents of project-planning mode.

Nonetheless, there are few notable exceptions, in which such team characteristics as stress, experience, empowerment, transactive memory system (TMS), trust, functional diversity, and demographic diversity were explicitly studied as antecedents of intuitive and political decision-making (e.g., Dayan & Di Benedetto, 2011; Dayan & Elbanna, 2011; Dayan et al., 2012). Yet, research designed to identify additional team-level variables that are of particular relevance to project-planning mode might yield additionally valuable insights on project outcomes, especially via employment of uncommon outcome measures. Given the above, there is a need for additional research that explicitly focuses on project-planning mode and its relationship to various team characteristics and project outcomes, which the present study sought to fulfill. Our study was informed by another concern on most previous studies which neglected the multifaceted nature of project performance (Elbanna, 2015; Elbanna et al., 2013). We remedy this limitation by including multiple measures that capture various aspects of project performance, namely, speed of project completion and project disturbance, and examining the nature of relationship between these aspects.

The rest of the study is organized in the following manner. First, the research model is introduced and project-planning mode is discussed in detail, followed by the development of the study hypotheses. Next, the study's methodology, data analysis, and findings are provided. Finally, the study findings are discussed, including a conclusion, and listing of the study's limitations and implications for managers and researchers.

11.2 Theory and Hypotheses

The central argument in this study is that the project-planning mode (emergent vs. deliberate) chosen for any given project is affected by various project team characteristics (project management skills and team stability) and that project-planning mode affects project outcomes (speed of project completion and project disturbance). Figure 11.1 depicts these relationships, along with the three control variables (industry type, company size, and company age) employed to prevent their possible confounding effects on the relationships portrayed.

11.3 Project-Planning Mode

The primary purpose of project planning is to divide a project into manageable elements or tasks that can be effectively managed. Among the tasks associated with project planning are selecting the project team, identifying team member roles and responsibilities, identifying the deliverables and creating a work breakdown structure, identifying the activities needed to complete the deliverables, and risk planning (Kerzner, 2013). The literature suggests that these and the like tasks of project

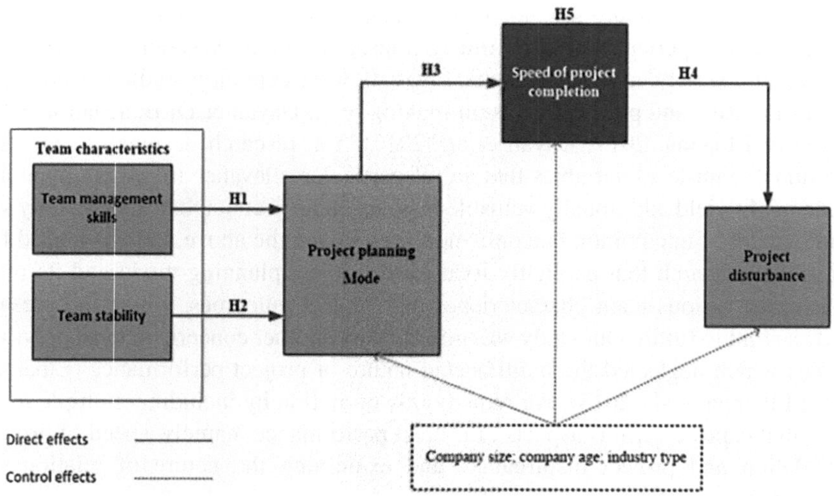

Fig. 11.1 Project-planning mode and its relationship to team characteristics and project outcomes

planning can be carried out in several ways, varying from a purely emergent to a purely deliberate mode. A purely emergent mode refers to when project planning tasks are developed without explicit intentions, whereas a purely deliberate mode refers to when the tasks are established based on precise intentions that are articulated in detail in advance, are known to all organizational members, and are realized exactly as intended without the interference of external forces (Mintzberg & Waters, 1985).

The main difference between the two modes is that the emergent mode produces evolving patterns via which project planning tasks are performed in an incremental piece-by-piece manner, whereas the deliberate mode produces patterns from imposed intention, directions, and control, via which tasks are completed in an analytical, systematic, and intentional manner (Ghobadian et al., 2008). Nonetheless, use of the emergent mode does not mean chaos or that management has no role in developing plans; it simply means that the process of developing plans is flexible and responsive. The deliberate mode, however, is used to obtain specific outcomes via direction and control (Mintzberg & Waters, 1985).

11.4 Team Characteristics as Antecedents of Project Planning Mode

We think that the planning mode chosen for use on any project is affected by various team characteristics. The present study investigated the role of two such important characteristics: team project management skills and team stability (Akgün & Lynn, 2002; Dayan & Colak, 2008; Wu et al., 2019). Below is a discussion of these variables as related to the present study, including the justification for their selection.

11.4.1 Team Management Skills and Project-planning Mode

As described by Hoegl and Parboteeah (2006), team project management skills was defined (subsequently referred to as team management skills) as the ability of project team members to plan a project (i.e., the ability to define work requirements and the resources needed, to divide a project into manageable sub-components, to identify sub-goals, and to organize workflow) and control its progress (i.e., the ability to track its progress, to compare actual outcome to predicted outcome, and to analyze impact) (Gladstein, 1984; Project Management Institute, 2013).

As compared to the emergent mode of planning, the deliberate mode involves a higher level of planning and more extensive analysis of data, greater reliance on formal planning tools, explicit formalized documentation and procedures, more frequent planning sessions and assessments, greater reliance on assessment of alternatives, more formal and informal communication, and more team member interaction (Dvir et al., 2003; Gibson & Gebken, 2003; Yu et al., 2007). These characteristics indicate that the deliberate mode of planning is more challenging and demanding, requiring a higher level of team management skills than is necessary when using the emergent mode of planning. This is because team management skills may be particularly necessary for effectively addressing the above-mentioned and similar planning activates associated with this mode (e.g., creating detailed and explicit documentations, planning the workflow, monitoring project's progress, using planning tools effectively, etc.).

These arguments suggest that the choice between the two planning modes may depend on the level of team management skills. More specifically, a project team with high level of management skills might have the level of confidence in its ability to effectively meet the requirements demanded by deliberate planning and thus may tend to choose the deliberate planning mode for any given project.

Research on self-efficacy and potency theories provides support for the notion that team member confidence in or perceptions of their management skills affect the choice of planning mode. More specifically, self-efficacy theory argues that an individual's self-efficacy (an individual's belief in his/her own capabilities) is an important determinant of his/her behavior and decisions (Bandura, 1995). In terms of project management, the self-efficacy of project team members is associated with team management skills (Blomquist et al., 2016).

Based on early findings that individuals with a high level of self-efficacy are more motivated and exert greater effort to complete challenging tasks, and generate more effective task strategies (Paunonen & Hong, 2010), we think that project teams consisting of members with a high level of self-efficacy are likely to be more confident in their ability to handle the challenging demands of the deliberate mode of planning and choose this planning mode for any given project.

Similarly, potency theory suggests that team members' belief in their ability to effectively perform as a team (team potency) (Guzzo, 1986) can affect their choice of planning mode. In particular, team members' perceptions of their team's potency may result in team members choosing the deliberate mode of planning.

In total, the above arguments on self-efficacy and team potency indicate that there is a relationship between team management skills and project-planning mode. In particular, teams with a high level of team management skills may be inclined to employ the deliberate mode of planning due to the members' perceptions of a high level of self-efficacy and/or team potency. More formally stated:

H1: A high level of team management skills is positively related to the use of the deliberate mode of project planning.

11.4.2 Team Stability and Project-planning Mode

In accordance to relevant research (e.g., Akgün & Lynn, 2002; Dayan & Colak, 2008), the present study defined team stability as the extent to which core team members and project managers remain on a project team until project completion. This variable is especially relevant to the choice of planning mode due to the nature of project teams; e.g., members may come from different functional units, may not have previously worked together, and are given time-limited and non-repetitive tasks that involve considerable application of knowledge, judgment, and expertise. Furthermore, building a team identity and effectively functioning as a team takes time (Cohen & Bailey, 1997; Savelsbergh et al., 2015).

It is known that team stability is related to various positive team or project-level outcomes, including team member relationships, the ability of each team member to correctly understand each other's level of knowledge, skill, and ability—so that tasks can be effectively allocated, team spirit, reliable planning, team learning, and project success (Akgün & Lynn, 2002; van Woerkom & Croon, 2009). Similarly, it was suggested that there is a direct relationship between team (in)stability and project-planning mode. Gibson and Gebken (2003), for example, posited that a high team member turnover rate (team instability) can decrease a team's ability to engage in intensive effective pre-project planning sessions, which is a part of the deliberate mode characterized by identification and analysis of project risks, formulation of early designs, critical decision-making, and definition of the specific project execution approach. Effective pre-project planning requires dynamic interactions between team members (Oertel & Antoni, 2015), which, in turn, requires a high level of formal and informal communication (Akgün et al., 2005; Gibson & Gebken, 2003; Yu et al., 2007).

For teams with a high turnover rate establishing or maintaining the required dynamic interactions and high level of communication, which are important for enhancing team TMS (an awareness of what each team member knows (Akgün et al., 2006) and development of team behavior, can be challenging. Team membership fluctuation (Bangerter, 2002) often has a negative effect on communication between team members (Akgün et al., 2005), which negatively affects team performance making it more difficult for members of such teams to develop a shared understanding of each other's working styles and knowledge, skills, and abilities due to interpersonal discontinuity of experience (Bangerter, 2002). Unstable teams are likely to be characterized by poor TMS (Lewis et al., 2007).

Even in cases in which planning activities are well-documented in advance (i.e., the deliberate mode), poor TMS and intragroup processes that are the result of unstable team membership can, over time, lead a project team to adapt emergent planning. Such change in the planning mode might be due to changes in a team's perceptions of self-efficacy or team potency regarding team management skills, the team's ability to conduct effective planning sessions, and/or the team's ability to work together effectively. In contrast, as a recent empirical study on knowledge management in IT-enabled business projects reported, stable teams are less likely to suffer from shifts in team member's perceptions of self-efficacy and are more likely to maintain use of the deliberate project-planning mode, which facilitates development of common goals and a shared understanding among team members (Gemino et al., 2015).

The arguments described above collectively indicate that team stability can be quite important to the deliberate mode of planning. According to group dynamics theory, team cohesion positively affects team performance (Beal et al., 2003; Dion, 2000). A closer review of previous studies supports this assertion. It was reported that stable teams have a higher level of shared experience with their tasks and teammates, have better quality communication, and are more effective at identifying specialized knowledge and coordinating expertise within the team, as compared to unstable teams (Cramton, 2001; Faraj & Sproull, 2000). Membership fluctuation can adversely affect any or a combination of these qualities or abilities of stable teams (Akgün et al., 2005; Bangerter, 2002); therefore, it is logical to expect that members of stable project teams work better together and are more likely to effectively use the deliberate mode of planning than unstable teams. More formally stated:

H2: The level of team stability is positively related to the use of the deliberate mode of project planning.

11.5 Speed of Project Completion and Project Disturbance

Research on future trends in business highlight that globalization, privatization, digitization, and rapid development of technology will create a hyper-competitive environment that will require businesses to operate at higher speed and more economically (Thamhain, 2009). To remain competitive in such an environment, businesses must be able to process and exchange information efficiently, as factors that affect decision-making change quickly. These expected characteristics of the future business environment suggest that the environment will also be characterized by an increased level of complexity and unexpected risk. These changes are also likely to challenge the use of traditional measures of project outcome, such as cost, schedule, and quality; therefore, businesses (and researchers) need to devise new outcome measures that are relevant to the requirements of the evolving business environment. As such, the present study used speed of project completion and project disturbance as measures of project outcome. Additionally, speed of project completion and project disturbance were used as outcome measures in response to some studies that highlight the multidimensional nature of project performance and suggest adoption of novel measures of outcome that are more holistic than those commonly used (Elbanna, 2018; Serrador & Turner, 2015a).

In accordance with the literature, in the present study, speed of project completion was defined as the extent to which a project is completed according to schedule (Dayan et al., 2012; Lynn et al., 2000) and project disturbance was defined as the extent to which the project yielded major negative unexpected outcomes (Elbanna et al., 2013; Rodrigues & Hickson, 1995). We think that the two study project outcome variables should be considered sequential or consequential variables, rather than distinct and unrelated. We provide evidence below to show that project-planning mode affects speed of project completion (H3) and that speed of project completion affects project disturbance (H4). The benefit of treating these two outcome variables as sequential is that doing so improves our understanding of how planning mode affects diverse project outcomes and facilitates comprehension of the interaction between project outcomes.

11.5.1 Project-planning Mode and Speed of Project Completion

It is widely argued that the quality of project planning positively affects project outcomes (Dvir & Lechler, 2004; Rudd et al., 2008). More specifically, it was proposed that project-planning mode is associated with speed of project completion and the deliberate mode of planning provides team members with a clearly defined purpose indicative of top management's intentions and an unambiguous sense of direction for completing all project sub-tasks. Consequently, project team members can understand and identify what the project requires of them and work toward completing the project with minimal deviation from the intended objectives (Mintzberg & Waters, 1985), resulting in timely completion. Of the two planning modes examined in the present study, the deliberate mode is regarded as the more useful mode, with the greatest potential for yielding favorable outcomes (Titus et al., 2010). In accordance with normative decision theory (Steele & Stefánsson, 2015), we propose that firms that adopt the deliberate mode of project planning will achieve better project outcomes than those using the emergent mode (Elbanna & Elsharnouby, 2018; Walter et al., 2012).

Earlier empirical research highlights the benefits of the deliberate mode of project planning and supports the above-mentioned proposal. For instance, studies on new product development (NPD) teams show that a clear strategy makes it easy for middle-level managers to prepare execution plans that consist of prioritizing work programs and budgets (Akgün & Lynn, 2002; Dayan & Colak, 2008). This is because a high degree of planning, which is characteristic of the deliberate mode, conceptually prepares project team members, focusing their attention on relevant opportunities for implementation and facilitating completion of projects on or ahead of schedule. Similarly, Gemino et al. (2015) studied 212 project managers of IT-enabled business projects and reported that explicit documentation, which is a characteristic of the deliberate mode of planning, helps team members understand each other's roles and to develop common goals that enable them to assess the progress of a project. Such teamwork and assessment have the potential to decrease the amount of time needed to complete a project.

The deliberate mode of project planning can help team members link project sub-tasks and quickly move from one sub-task to the next, resulting in timely project completion. In this vein, Marks et al. (2001) suggest that when team members use, or rely more on, the deliberate mode of project planning, they are able to complete projects faster than when using the emergent mode because deliberate planning reduces the frequency and extent of plan changes, which permits team members to smoothly move through all project sub-tasks. More formally stated:

H3: The deliberate mode of project planning is positively related to speed of project completion.

11.5.2 Project Completion Speed and Project Disturbance

Previous research suggests that as the speed of project completion increases, the likelihood of project disturbance (a project generates unexpected and unfavorable outcomes) decreases. More specifically, speed generates various positive organizational outcomes, including entrepreneurial discovery (Keupp & Gassmann, 2009), efficient introduction of new products and processes (Jones et al., 2001), high-quality decision implementation (Elbanna et al., 2014), and improved firm performance (Baum & Wally, 2003; Judge & Miller, 1991). Similarly, studies on NPD teams have shown that speed is associated with various dimensions of new product success, based on operational, customer-based, and financial measures (see, Cankurtaran et al., 2013). These positive outcomes arise because timely action enables firms to benefit from first-mover advantages (Makadok, 1998), build new competencies (Wheelwright & Clark, 1992), capitalize on opportunities before they disappear (Baum & Wally, 2003), and utilize a fast-follower strategy (Emmanuelides, 1993).

Similarly, we propose that speed of project completion is associated with various less-commonly examined positive outcomes at the project level. One such outcome is project disturbance. Project disturbance can be related to the resources or information allocated to a project (additional resources had to be allocated unexpectedly), to the quality of the interactions between all individuals involved in a project (Rodrigues & Hickson, 1995), and to the level of satisfaction with a completed project among its owners (e.g., project managers, customers, etc.). As the time to completion of a project increases, the potential for unexpected outcomes increases; therefore, as the speed of project completion increases the risk of project disturbance should decrease. More formally stated:

H4: Speed of project completion is positively related to a decrease in the risk of project disturbance.

11.6 Speed of Project Completion as a Mediator

Considering key intervening variables when examining planning-outcomes linkages can contribute to explain inconsistent results in the literature (Dibrell et al., 2014). As stated before, the present study examined two project-level outcome variables (speed of project completion and project disturbance) and considered them as sequential outcomes of planning mode. This required moving beyond the earlier postulated direct effects of planning mode on speed of project completion (H3) and those of speed of project completion on project disturbance (H4) to consideration of their joint interactions. H3 and H4 jointly suggest that the deliberate mode of project planning can exert its influence on project disturbance via speed of project completion or that speed of project completion serves as a mediator between planning mode and project disturbance by transmitting the former's effect to the latter (the broken line in Fig. 11.1).

Following Rungtusanatham et al. (2014), who recommend that any such mediating effects be explicitly stated, H5 was formulated. This mediation hypothesis describes a reasonable expectation that a well-planned project is likely to be completed on time and decrease the likelihood of project disturbance, such as delays and penalties. Furthermore, H5 facilitates a better understanding of the link between project-planning mode and project disturbance, showing the mechanism by which the former variable affects the latter.

H5: Speed of project completion mediates the relationship between the mode of project planning and project disturbance.

11.7 Methods

Data were collected from businesses in the UAE using a study-specific questionnaire, and the following four criteria were used to select study projects: (1) The project had to be recent (executed no more than 2 years earlier); (2) Project outcomes had to be known at the time of data collection; (3) Those completing the study questionnaire had to be closely involved in the selected project; and (4) The project had to be very important. As English is the primary language used by businesses in the UAE (Elbanna & Elsharnouby, 2018), the study questionnaire was developed in English. A draft of the questionnaire was reviewed by five academicians before administering the modified version to two professionals for feedback.

11.8 Data Collection

The study questionnaire was sent to 500 private companies located in the three largest Emirates in the UAE (Dubai, Abu Dhabi, and Sharjah) with <500 employees. The 410 companies that responded were included in the study. Questionnaires were completed by multiple responders at 40 of the companies, so as to evaluate inter-rater reliability. When contacting informants for appointments to collect data, we confirmed that their participation is voluntary.

The mean number of employees in the sampled companies was 120. The percentage of Emirati employees was very low (7%), as compared to that of foreigners (93%). In all, 58% of the sampled companies were in service industries and 42% were in manufacturing industries. Mean age of the sampled companies was 16 years, 78% were foreign owned, and 15% and 7% were locally owned and jointly owned (local and foreign co-owned), respectively. The sample incorporated a wide range of industries, including construction, IT, manufacturing, import-export, general trading, and finance. The database used for analysis consisted of 410 projects. The projects were also diverse and included entering a new market, introduction of a new product, installation of new IT systems, opening a new branch, adding new production facilities, and restructuring.

Most of the respondents were male (92%), whereas 8% were female. Most respondents were department managers (64%), followed by CEOs, presidents, and general managers (27%), branch, project, and area managers (7%), and not specified (2%). Mean age of the respondents was 42 years. In total, 68% of the respondents had a university degree, 28% had a graduate degree, and 4% had a high school degree.

11.9 Measures

This study assesses all measures at the project-level because complexity, size, and uniqueness of projects vary (Mir & Pinnington, 2014); and hence, within the same company, the processes and consequences of projects are expected to differ (Müller & Turner, 2007). With three exceptions, company size (number of employees), company age (number of years since establishment), and industry type (a dummy variable), all constructs were measured using Likert five-point scale questions.

We measured project-planning mode across five items scales derived from Titus et al. (2010). These are "(1) we typically knew what the content of the plan of this project should be without engaging in trial-and-error actions. (2) The plan of this project was typically prepared in advance, rather than emerging over time as the best means for achieving our objectives become clearer. (3) Our project was carefully planned and well understood before any significant actions were taken. (4) A formal plan served as the basis for our actions in this project. (5) The plan of this project typically resulted from a formal planning process (i.e. the formal plan precedes the action)." Conducting confirmatory factor analysis (CFA) led to removing the first item from the scale.

Using three-item scales adopted from related research (Dayan et al., 2012; Lynn et al., 2000), the speed of project completion was operationalized. These are "(1) this project was completed in less time than what was considered normal and customary for our firm. (2) This project was completed on or ahead of the original schedule developed at initial project go-ahead. (3) Top management was pleased with the time it took us to implement this project." To measure project disturbance, we followed Elbanna et al. (2013) and asked respondents: "to what extent were there any major negative unexpected outcomes of this project?"

Team management skills were measured by using two-item scales of Hoegl and Parboteeah (2006). These are "(1) the team had the necessary skills for planning the project (e.g., setting sub-goals, organizing the work, etc.). (2) The team had the necessary skills to control the progress of the project." To measure team stability, three-item scales adapted from related research (Akgün & Lynn, 2002; Dayan & Colak, 2008) were used. These are "(1) the team leader who started this project remained on it from the beginning of the project through its end. (2) Team members who were on the team remained on it from the beginning of the project through its end. (3) There were no major changes in the team members during this project." Finally, we controlled for the effects of three variables, namely, company size, company age, and industry type, because past research suggests that these variables may affect project processes and outcomes (e.g., Dayan & Elbanna, 2011; Müller & Turner, 2007).

11.10 Reliability, Validity, and Bias

Since the coefficients of Cronbach α ranged from 0.71 to 0.86, we claim that the study variables have a satisfactory degree of internal consistency (please, see Table 11.1). The relatively low to moderate correlations between variables as shown in Table 11.1 (the highest correlation between variables is 0.53) provide evidence of discriminant validity (Hair et al., 2009). To assess our scales, we conducted CFA. As shown in Table 11.2, the measurement model produced a good fit with GFI of 0.94; AGFI of 0.89; CFI of 0.93; IFI of 0.93; RMSEA of 0.08; and SRMR of 0.05. Other fit indices such as $\chi 2 = 176.07$ with $df = 45$, $\chi 2/df = 3.91$ are also acceptable (Hair et al., 2009). Moreover, all item measures had SFLs (standardized factor loadings) which well exceeded the recommended threshold of 0.50 (ranging between 0.60 and 0.91) (Hair et al., 2009). These results show acceptable unidimensionality and demonstrate that the measurement model is satisfactory.

Table 11.1 Descriptive statistics and reliability estimates ($N = 410$)

Variables	Mean	SD	Alpha	A.	B.	C.	D.	E.	F.	G.
A. Project disturbance	2.52	0.91	NA	1						
B. Speed of project completion	3.31	0.75	0.71	−0.14**	1					
C. Project-planning mode	3.78	0.70	0.78	−0.14**	0.25***	1				
D. Team management skills	3.85	0.86	086	−0.15**	0.33***	0.53***	1			
E. Team stability	3.73	0.78	0.79	−0.13**	0.22***	0.45***	0.49***	1		
F. Company size	119	89	NA	−0.10*	−0.05	0.07	0.04	0.06	1	
G. Company age	16	9	NA	0.03	0.10^{+}	0.02	0.09^{+}	0.14**	0.24***	1
H. Industry type	1.59	0.49	NA	0.01	−0.04	0.15**	0.13*	0.07	0.09^{+}	−0.13**

$^{+}P < 0.10$; $^{*}P < 0.05$; $^{**}P < 0.01$; $^{***}P < 0.001$; *NA* not applicable

Table 11.2 Goodness-of-fit indices

Model	$\chi2$	df	$\chi2$/df	GFI	AGFI	CFI	IFI	RMSEA	SRMR	Confidence interval
Measurement model	176.07	45	3.91	0.94	0.89	0.93	0.93	0.08	0.05	0.07–0.10
Structural model (restricted)	265.20	82	3.23	0.93	0.88	0.91	0.91	0.07	0.05	0.06–0.08
Structural model (full)	260.21	81	3.21	0.93	0.88	0.91	0.91	0.07	0.05	0.06–0.08

Df degrees of freedom, *GFI* goodness-of-fit index, *AGFI* adjusted goodness-of-fit index, *CFI* comparative fit index, *IFI* incremental fit index, *RMSEA* root mean squared error of approximation, *SRMR* standardized root mean squared residual

Because in 90% of the examined projects data was collected from single respondents, we attempted to reduce bias, such as retrospective rationalization and incomplete recall. This was done as follows. Before data collection, we (1) examined recent projects only which had clear outcomes at the time of collecting data; (2) assured informants of confidentiality; (3) reversed scale anchors in some places; (4) measured company size, company age, and industry type using objective data; (5) ordering the survey questions in a way to prevent respondents from guessing examined relationships; and (6) tested relationships which informant were unable to predict, such as mediation effects.

After data collection, we examined the possibility of bias using two tests. First, the 40 companies with two informants demonstrated significant correlations at either the 5% level (two companies) or better (38 companies) with an average of correlation coefficients of 0.61 showing a strong level of inter-rater reliability. Second, to assess common method bias, CFA was applied to single-factor model (Sanchez & Brock, 1996). The model fit indices of $\chi2$ (54) = 799.59, GFI = 0.77, AGFI = 0.67, CFI = 0.65, IFI = 0.65, RMSEA = 0.18, and SRMR = 0.11 were unacceptable and significantly worse than those of the measurement model. This proposes that a single-factor model is not acceptable; thus, any potential common method bias is not significant.

11.11 Results

Table 11.1 shows correlations and descriptive statistics for the variables in our conceptual model. The correlation coefficients do not suggest multicollinearity concerns since they are all below 0.55. We began the data analysis by performing a CFA as discussed above which shows that our measurement model has a good fit. In the second stage of data analysis, the structural relations among the variables in our conceptual model including the control variables were tested with path analysis, in two steps, using the maximum likelihood estimation technique in SPSS AMOS

Table 11.3 Standardized parameter estimates ($n = 410$)

	Path to					
	Restricted model (RM)			Full model (FM)		
Path from	Project-planning mode (RM 1)	Speed of project completion (RM 2)	Project disturbance (RM 3)	Project-planning mode (FM 1)	Speed of project completion (FM 2)	Project disturbance (FM 3)
Company size	0.06	−0.10+	−0.10*	0.06	−0.10⁺	−0.12*
Industry type	0.07	−0.08	0.06	0.07	−0.08	0.04
Company age	−0.06	0.11⁺	0.06	−0.06	0.11⁺	0.08
Team management skills	0.53***			0.53***		
Team stability	0.24**			0.24**		
Project-planning mode		0.40***	−0.18**		0.39***	−0.12⁺
Speed of project completion						−0.14*
R^2	0.51	0.17	0.05	0.51	0.17	0.06

⁺$P < 0.10$; *$P < 0.05$; **$P < 0.01$; ***$P < 0.001$

version 18. First, to control for the effect of speed of project completion, we ran the path model on the sample without examining the path from speed of project completion to project disturbance (restricted model). Second, we reran the path model on the sample with the additional path between speed of project completion and project disturbance in order to examine the mediating effect of speed of project completion (full model).

To test how well the data fit both structural models, we used the traditional goodness-of-fit indices mentioned above. The restricted structural model produced a good fit with $\chi2 = 265.20$, df = 82; $\chi2/\text{df} = 3.23$; GFI of 0.93; AGFI of 0.88; CFI of 0.91; IFI of 0.91; RMSEA of 0.07; and SRMR of 0.05. Equally, the full structural model produced a good fit with $\chi2 = 260.21$, df = 81; $\chi2/\text{df} = 3.21$; GFI of 0.93; AGFI of 0.88; CFI of 0.91; IFI of 0.91; RMSEA of 0.07; and SRMR of 0.05. Finally, item measures in both models had all SFLs ranging between 0.63 and 0.91 that is above the recommended threshold of 0.60 (Hair et al., 2009). The above results show that both structural models are acceptable.

The results of restricted model were used to test the first three hypotheses. H1 related team management skills to the use of planned mode in planning projects. This hypothesis was supported with a β-coefficient of 0.53 ($P < 0.001$) (see RM 1 in Table 11.3). H2 suggested a positive relationship between team stability and the use of planned mode. Similarly, this relationship was supported by our findings with a

β-coefficient of 0.24 ($P < 0.01$) (see RM 1 in Table 11.3). H3, which suggested a positive relationship between the use of planned mode in planning projects and project completion speed, and was also supported with a β-coefficient of 0.40 ($P < 0.001$) (see RM 2 in Table 11.3).

We then used the results of full model to examine H4 as long as confirming the results of the first three hypotheses. The results of full model confirm our earlier results concerning the first three hypotheses. Specifically, both team management skills ($\beta = 53$, $P < 0.001$, see FM 1 in Table 11.3) and team stability ($\beta = 24$, $P < 0.01$, see FM 1 in Table 11.3) were positively related to the use of planned mode; while the use of planned mode had a positive effect on speed of project completion ($\beta = 39$, $P < 0.001$, see FM 2 in Table 11.3). H4 suggesting a negative impact of project completion speed on project disturbance was supported with a β-coefficient of −0.14 ($p < 0.05$, see FM 3 in Table 11.3).

Finally, H5, which postulated that project completion speed mediates the relationship between planned mode and project disturbance, was tested using the results of both restricted and full models according to the procedure suggested by Baron and Kenny (1986) who recommended four conditions for a variable to act as a mediator. These are (1) the predicting variable (project-planning mode) significantly influences the mediator variable (speed of project completion). (2) Project-planning mode significantly influences project disturbance when we do not consider the impact of speed of project completion. (3) Speed of project completion significantly influences project disturbance. (4) When both project-planning mode and speed of project completion are considered together, the impact of project-planning mode (the predicting variable) on project disturbance (the dependent variable) is either reduced (partial mediation) or no longer significant (full mediation) because of the possible impact of speed of project completion (the mediator).

The first two conditions are examined using the restricted model; while the last two conditions are tested using the full model. As revealed above, the first condition was achieved, since project-planning mode significantly influences speed of project completion ($\beta = 40$, $P < 0.001$; RM 2 in Table 11.3). Similarly, project-planning mode significantly influences project disturbance ($\beta = -0.18$, $P < 0.01$; RM 3 in Table 11.3), which meets the second condition of a mediating effect. The results of full model provide support to the last two conditions. Specifically, speed of project completion negatively influences project disturbance ($\beta = -0.14$, $P < 0.05$) and the impact of project-planning mode on project disturbance is reduced ($\beta = -0.12$, $P < 0.10$) when both planning mode and speed of project completion are incorporated in the same model (FM 3 in Table 11.3) compared to its explanatory value in the restricted model ($\beta = -0.18$, $P < 0.01$, RM 3 in Table 11.3). These results show a partial mediation effect of speed of project completion on the relationship between planning mode and project disturbance, and hence, H5 was supported.

11.12 Discussion and Conclusion

11.12.1 Interpretation

The present findings support the study's central hypothesis—that the project-planning mode chosen for any given project is affected by team management skills (H1) and team stability (H2). Our findings on this central argument can be linked to those of early studies in team settings. More specifically, the finding that team management skills (H1) and team stability (H2) are positively related to deliberate mode of project planning indicates the importance of these team characteristics to team potency, TMS, and self-efficacy, which is in agreement with earlier studies (see part 2.2.1 and part 2.2.2) that reported various positive outcomes of these variables. A better understanding of these outcome variables could benefit team performance and overall performance of organizations that rely heavily on teams (see also, Chen et al., 2009; Collins & Parker, 2010; Mathieu et al., 2008).

Another important finding of the present study is that the deliberate mode of planning is positively correlated with speed of project completion (H3) and a decrease in the risk of project disturbance (H4). The present finding that the deliberate mode of planning is positively related to speed of project completion (H3) is in agreement with earlier findings showing that intuitive decision-making—a departure from the deliberate mode—is negatively associated with decision-making speed (Dane & Pratt, 2009) and organizational performance (Khatri & Ng, 2000), that procedural rationality (i.e., extensively analyzing the relevant information before making strategic decisions) is positively related to the successful implementation and outcomes of strategic decisions (Al-Hashimi et al., 2022), and that process formalization (i.e., the use of explicit rules and standard procedures), a characteristic of the deliberate mode, improves coordination of project tasks, reduces uncertainty, and expedites NPD (Li & Atuahene-Gima, 1999).

Similarly, the present finding that speed of project completion is negatively related to project disturbance (H4) is also consistent with early studies that show that there are various positive outcomes of NPD speed, such as product success, marketplace success, and operational efficiency (Cankurtaran et al., 2013). Completing projects without delays and major unexpected outcomes is especially important in today's competitive business environment, in which time-based strategies are increasingly recognized to be a competitive weapon (Stalk, 1990). This is especially desirable when multiple projects (e.g., NPD projects) are undertaken simultaneously since delay in the completion of a project can cause delays in other projects, preventing an organization from benefiting from a particular project and other projects at hand, leading to unforeseen large-scale negative consequences. This butterfly effect can occur due to the dependence of multiple projects on the same resources and strategic commitment provided by an organization (Krohwinkel, 2015). The present findings related to H3 and H4 collectively suggest that the use of deliberate mode of project planning can help businesses avoid this butterfly effect.

The present study's finding related to H5, which shows that the deliberate mode of planning reduces the likelihood of project disturbance by facilitating timely project completion, implies that this mode of planning can help organizations avoid the adverse effects associated with delays in the completion of projects. This finding is consistent with earlier studies that reported decision disturbance is positively associated with intuitive decision-making and negatively associated with rational decision-making (Elbanna et al., 2013), which are considered variants of the emergent and deliberate modes of project planning, respectively.

11.12.2 Limitations and Future Research

The present study has some limitations, including measurement of only two project outcomes. Additional research that employs additional measures of project outcome, such as cost, quality, satisfaction, and project ending performance (An et al., 2019), is required to more clearly discern how team characteristics and planning mode affect project outcomes (Leybourne & Sadler-Smith, 2006). Another limitation is that the sample included companies doing business in varying types of industries and projects although the variables employed in the present study might have highlighted different characteristics across the industry or project types, and thus need to be controlled.

Another limitation is that the present study examined the effect of only two project team factors (team management skills and stability) on project-planning mode. Future studies should examine the role of a larger number of team-level variables, such as team diversity and team experience (Mathieu et al., 2008). It can also develop more comprehensive models that articulate how planning mode and project outcomes are affected by various contextual variables that operate at the project level, including project risk, project complexity, and project uncertainty (Jun et al., 2011; Padalkar & Gopinath, 2016), the organizational level, including strategic learning (Sirén & Kohtamäki, 2016), and the environmental level, including environmental dynamism and national culture (Shepherd & Rudd, 2014; Yousfi, 2019).

Project risk or the other contextual variables at different levels mentioned above (e.g., project complexity and environmental dynamism) can also be incorporated into the current study through a contingency theory approach, which suggests that the impact of project planning on project outcomes depends on contextual variables (Dvir & Lechler, 2004). The utility of contingency theory in project management research is becoming more popular (Hanisch & Wald, 2012; Shepherd & Rudd, 2014; Söderlund, 2011). Via this theoretical lens, researchers of project management will be able to respond to recent calls for employing broader conceptualizations and alternative theoretical perspectives (Floricel et al., 2014; Svejvig & Andersen, 2015) and yield a more complete understanding of effective project planning.

11.12.3 Managerial Implications

The present findings have important implications for managers, as they show that team stability and team management skills are significantly associated with the deliberate mode of project planning, indicating that project managers should carefully monitor team building and development, recruitment of qualified members, and team member turnover. To effectively do so, project managers can:

- use various practices that will mitigate the negative effects of team instability on team functioning and project outcomes, such as electronic knowledge repositories and succession planning (Pee et al., 2014);
- identify competency gaps and compose project teams with the most suitable members (Liu & Cross, 2016);
- employ effective mechanisms, such as use of a clear team structure (Bunderson & Boumgarden, 2010), participative leadership style (Burris et al., 2009), and well-designed incentive systems (Edmondson, 2004) to enhance team psychological wellbeing and ensure team stability (Chandrasekaran & Mishra, 2012); and
- provide team training designed to improve team management skills and stability (Edmondson, 2004; Goldstein & Ford, 2002; Salas et al., 2008).

When provided simultaneously (rather than in an isolated or sequential manner) and integrated into the early stages of project planning, the strategies described above facilitate team development and team members' acquisition of project management skills as an integrated process (Thomas et al., 2008), thus helping team members function more effectively, resulting in successful project planning and expected project outcomes.

In sum, the study's findings suggest from a practitioner perspective the followings:

- Team management skills and team stability influence the project-planning mode.
- Adopting a more deliberate mode in planning projects is more likely to succeed.
- As the speed of project completion increases, the risk of project disturbance should decrease.

Project managers should carefully monitor team building and development and team member turnover.

Acknowledgement The authors acknowledge the funds received to conduct this study from the National Research Foundation/the UAEU grant (grant # 101 NRF-UAEU: 3209-31B014) and the Coordination Office for Scientific Research Projects (BAP) of the Social Sciences University of Ankara (Project #: SBA-2017-123). The contributions of both authors are equal.

References

Akgün AE, Lynn GS (2002) New product development team improvisation and speed-to-market: an extended model. Eur J Innov Manag 5:117–129

Akgün AE, Byrne J, Keskin H, Lynn GS, Imamoglu SZ (2005) Knowledge networks in new product development projects: a transactive memory perspective. Inf Manag 42(8):1105–1120

Akgün AE, Byrne JC, Keskin H, Lynn GS (2006) Transactive memory system in new product development teams. IEEE Trans Eng Manag 53(1):95–111

Al-Hashimi K, Weerakkody V, Elbanna S, Schwarz G (2022) Strategic decision making and implementation in public organizations in the gulf cooperation council: the role of procedural rationality. Public Adm Rev 82(5):905–919

An N, Qiang M, Wen Q, Jiang H, Xia B (2019) Contribution of project managers' capability to project ending performance under stressful conditions. Eur Manag J 37(2):198–209

Bandura A (1995) Self-efficacy in changing societies. Cambridge University Press, New York

Bangerter A (2002) Maintaining interpersonal continuity in groups: The role of collective memory processes in redistributing information. Group Process Intergroup Relat 5(3):203–219

Baron RM, Kenny DA (1986) The moderator-mediator variable distinction in social psychological research: Conceptual, strategic, and statistical considerations. J Pers Soc Psychol 51(6):1173–1182

Baum JR, Wally S (2003) Strategic decision speed and firm performance. Strateg Manag J 24(11):1107–1129

Beal DJ, Cohen RR, Burke MJ, McLendon CL (2003) Cohesion and performance in groups: a meta-analytic clarification of construct relations. J Appl Psychol 88(6):989–1004

Blomquist T, Farashah AD, Thomas J (2016) Project management self-efficacy as a predictor of project performance: constructing and validating a domain-specific scale. Int J Proj Manag 34(8):1417–1432

Bunderson JS, Boumgarden P (2010) Structure and learning in self-managed teams: why "bureaucratic" teams can be better learners. Organ Sci 21(3):609–624

Burris ER, Rodgers MS, Mannix EA, Hendron MG, Oldroyd JB (2009) Playing favorites: the influence of leaders' inner circle on group processes and performance. Personal Soc Psychol Bull 35(9):1244–1257

Cankurtaran P, Langerak F, Griffin A (2013) Consequences of new product development speed: a meta-analysis. J Prod Innov Manag 30(3):465–486

Chandrasekaran A, Mishra A (2012) Task design, team context, and psychological safety: an empirical analysis of R&D projects in high technology organizations. Prod Oper Manag 21(6):977–996

Chen G, Kanfer R, DeShon RP, Mathieu JE, Kozlowski SWJ (2009) The motivating potential of teams: Test and extension of Chen and Kanfer's (2006) cross-level model of motivation in teams. Organ Behav Hum Decis Process 110(1):45–55

Cohen SG, Bailey DE (1997) What makes teams work: Group effectiveness research from the shop floor to the executive suite. J Manag 23(3):239–290

Collins CG, Parker SK (2010) Team capability beliefs over time: distinguishing between team potency, team outcome efficacy, and team process efficacy. J Occup Organ Psychol 83(4):1003–1023

Cramton CD (2001) The mutual knowledge problem and its consequences for dispersed collaboration. Organ Sci 12(3):346–371

Dane E, Pratt MG (2009) Conceptualizing and measuring intuition: A review of recent trends. In: Hodgkinson G, Ford JK (eds) International review of industrial and organizational psychology, vol 24. Wiley, Chichester, pp 1–40

Dayan M, Colak M (2008) The role of procedural justice in the new product development process. Eur J Innov Manag 11(2):199–218

Dayan M, Di Benedetto CA (2011) Team intuition as a continuum construct and new product creativity: the role of environmental turbulence, team experience, and stress. Res Policy 40:276–286
Dayan M, Elbanna S (2011) Antecedents of team intuition and its impact on the success of new product development projects. J Prod Innov Manag 28(S1):159–174
Dayan M, Elbanna S, Benedetto CAD (2012) Antecedents and consequences of political behavior in new product development teams. IEEE Trans Eng Manag 59(3):470–482
Dibrell C, Craig JB, Neubaum DO (2014) Linking the formal strategic planning process, planning flexibility, and innovativeness to firm performance. J Bus Res 67(9):2000–2007
Dion KL (2000) Group cohesion: from "field of forces" to multidimensional construct. Group Dyn Theory Res Pract 4(1):7–26
Dvir D, Lechler T (2004) Plans are nothing, changing plans is everything: the impact of changes on project success. Res Policy 33(1):1–15
Dvir D, Raz T, Shenhar AJ (2003) An empirical analysis of the relationship between project planning and project success. Int J Proj Manag 21(2):89–95
Edmondson AC (2004) Psychological safety, trust, and learning in organizations: a group-level lens. In: Kramer R, Cook K (eds) Trust and distrust in organizations: dilemmas and approaches. Russell Sage Foundation, New York, pp 239–272
Elbanna S (2015) Intuition in project management and missing links: analyzing the predicating effects of environment and the mediating role of reflexivity. Int J Proj Manag 33(6):1236–1248
Elbanna S (2018) The constructive aspect of political behavior in strategic decision-making: the role of diversity. Eur Manag J 36(5):616–626
Elbanna S, Elsharnouby TH (2018) Revisiting the relationship between formal planning process and planning effectiveness: Do organizational capabilities and decision-making style matter? Int J Contemp Hosp Manag 30(2):1016–1034
Elbanna S, Child J, Dayan M (2013) A model of antecedents and consequences of intuition in strategic decision-making: evidence from Egypt. Long Range Plan 46(1–2):149–176
Elbanna S, Thanos IC, Colak M (2014) An exploratory study of the determinants of the quality of strategic decision implementation in Turkish industrial firms. J Gen Manag 40(2):27–46
Emmanuelides PA (1993) Towards an integrative framework of performance in product development projects. J Eng Technol Manag 10(4):363–392
Faraj S, Sproull L (2000) Coordinating expertise in software development teams. Manag Sci 46(12):1554–1568
Floricel S, Bonneau C, Aubry M, Sergi V (2014) Extending project management research: insights from social theories. Int J Proj Manag 32(7):1091–1107
Gemino A, Reich BH, Sauer C (2015) Plans versus people: comparing knowledge management approaches in IT-enabled business projects. Int J Proj Manag 33(2):299–310
Ghobadian A, O'Regan N, Thomas H, Liu J (2008) Formal strategic planning, operating environment, size, sector and performance: evidence from the UK's manufacturing SMEs. J Gen Manag 34(2):1–20
Gibson E, Gebken R (2003) Design quality in pre-project planning: applications of the project definition rating index. Building Research and Information 31(5):346–356
Gladstein DL (1984) Groups in context: a model of task group effectiveness. Adm Sci Q 29:499–517
Goldstein IL, Ford JK (2002) Training in organizations: needs assessment, development, and evaluation. Wadsworth, Belmont, CA
Guo J, Zhou B, Zhang H, Hu C, Song M (2020) Does strategic planning help firms translate slack resources into better performance? J Manag Organ 26(4):395–407
Guzzo RA (1986) Group decision making and group effectiveness in organizations. In: Goodman P (ed) Designing effective work groups. Jossey-Bass, San Francisco, pp 34–71
Hair JF, Black WC, Babin BJ, Anderson RE (2009) Multivariate data analysis, 7th edn. Prentice Hall, Englewood Cliffs, NJ
Hanisch B, Wald A (2012) A bibliometric view on the use of contingency theory in project management research. Proj Manag J 43(3):4–23

Hoegl M, Parboteeah KP (2006) Team reflexivity in innovative projects. R&D Manag 36(2):113–125
Jones GK, Lanctot A, Teegen HJ (2001) Determinants and performance impacts of external technology acquisition. J Bus Ventur 16(3):255–283
Judge WQ, Miller A (1991) Antecedents and outcomes of decision speed in differential environmental context. Acad Manag J 34(2):449–463
Jun L, Qiuzhen W, Qingguo M (2011) The effects of project uncertainty and risk management on IS development project performance: a vendor perspective. Int J Proj Manag 29(7):923–933
Kerzner H (2013) Project management: a systems approach to planning, scheduling, and controlling. Wiley, Hoboken, NJ
Keupp MM, Gassmann O (2009) The past and the future of international entrepreneurship: a review and suggestions for developing the field. J Manag 35(3):600–633
Khatri N, Ng HA (2000) The role of intuition in strategic decision making. Hum Relat 53(1):57–86
Krohwinkel A (2015) A configurational approach to project delays: evidence from a sequential mixed methods study. J Mixed Methods Res 9(4):335–361
Lewis K, Belliveau M, Herndon B, Keller J (2007) Group cognition, membership change, and performance: Investigating the benefits and detriments of collective knowledge. Organ Behav Hum Decis Process 103(2):159–178
Leybourne S, Sadler-Smith E (2006) The role of intuition and improvisation in project management. Int J Proj Manag 24(6):483–492
Li H, Atuahene-Gima K (1999) Marketing's influence and new product performance in Chinese firms. J Int Mark 7(1):34–56
Liu W-H, Cross JA (2016) A comprehensive model of project team technical performance. Int J Proj Manag 34(7):1150–1166
Lynn GS, Reilly RR, Akgün AE (2000) Knowledge management in new product teams: practices and outcomes. IEEE Trans Eng Manag 47:221–231
Makadok RC (1998) Can first-mover and early-mover advantages be sustained in an industry with low barriers to entry/imitation? Strateg Manag J 19(7):683–696
Marks MA, Mathieu JE, Zaccaro SJ (2001) A temporally based framework and taxonomy of team processes. Acad Manag Rev 26(3):356–376
Mathieu J, Maynard MT, Rapp T, Gilson L (2008) Team effectiveness 1997-2007: a review of recent advancements and a glimpse into the future. J Manag 34(3):410–476
Mintzberg H, Waters JA (1985) Of strategies, deliberate and emergent. Strateg Manag J 6(3):257–272
Mir FA, Pinnington AH (2014) Exploring the value of project management: linking project management performance and project success. Int J Proj Manag 32(2):202–217
Müller R, Turner R (2007) The influence of project managers on project success criteria and project success by type of project. Eur Manag J 25(4):298–309
O'Shannassy TF (2017) Associate editor reflections on the progress in and future of strategic management research. *Journal of Management and Organization* 23(4):473–482
Oertel R, Antoni CH (2015) Phase-specific relationships between team learning processes and transactive memory development. Eur J Work Organ Psy 24(5):726–741
Padalkar M, Gopinath S (2016) Are complexity and uncertainty distinct concepts in project management? A taxonomical examination from literature. Int J Proj Manag 34(4):688–700
Paunonen SV, Hong RY (2010) Self-efficacy and the prediction of domain-specific cognitive abilities. J Pers 78(1):339–360
Pee LG, Kankanhalli A, Tan GW, Tham G (2014) Mitigating the impact of member turnover in information systems development projects. IEEE Trans Eng Manag 61(4):702–716
Project Management Institute (2013) A guide to the project management body of knowledge (PMBOK® Guide), 5th edn. Project Management Institute, Newtown Square, PA
Rodrigues SB, Hickson DJ (1995) Success in decision making: different organizations, differing reasons for success. J Manag Stud 32(5):655–678
Rudd JM, Greenley GE, Beatson AT, Lings IN (2008) Strategic planning and performance: extending the debate. J Bus Res 61(2):99–108

Rungtusanatham M, Miller JW, Boyer KK (2014) Theorizing, testing, and concluding for mediation in SCM research: tutorial and procedural recommendations. J Oper Manag 32(3):99–113
Salas E, DiazGranados D, Klein C, Burke CS, Stagl KC, Goodwin GF, Halpin SM (2008) Does team training improve team performance? A meta-analysis. *Human Factors* 50(6):903–933
Sanchez JI, Brock P (1996) Outcomes of perceived discrimination among Hispanic employees: Is diversity management a luxury or a necessity? Acad Manag J 39(3):704–719
Savelsbergh CMJH, Poell RF, van der Heijden BIJM (2015) Does team stability mediate the relationship between leadership and team learning? An empirical study among Dutch project teams. Int J Proj Manag 33(2):406–418
Serrador P, Turner R (2015a) The relationship between project success and project efficiency. Proj Manag J 46(1):30–39
Serrador P, Turner R (2015b) What is enough planning? Results from a global quantitative study. IEEE Trans Eng Manag 62(4):462–474
Shepherd NG, Rudd JM (2014) The Influence of context on the strategic decision-making process: a review of the literature. Int J Manag Rev 16(3):340–364
Sirén C, Kohtamäki M (2016) Stretching strategic learning to the limit: the interaction between strategic planning and learning. J Bus Res 69(2):653–663
Söderlund J (2011) Pluralism in project management: navigating the crossroads of specialization and fragmentation. Int J Manag Rev 13(2):153–176
Stalk G (1990) Competing against time: how time-based competition is reshaping global markets. Simon and Schuster, New York
Steele K, Stefánsson HO (2015) Decision theory. In: Zalta EN (ed) Stanford encyclopedia of philosophy. Metaphysics Research Lab, CSLI, Stanford University, Stanford
Svejvig P, Andersen P (2015) Rethinking project management: A structured literature review with a critical look at the brave new world. Int J Proj Manag 33(2):278–290
Thamhain HJ (2009) The future of project team leadership in complex project environments. In: Cleland D, Bidanda B (eds) Project management circa 2025. Project Management Institute, Newtown Square, PA, pp 365–382
Thomas M, Jacques PH, Adams JR, Kihneman-Wooten J (2008) Developing an effective project: Planning and team building combined. Proj Manag J 39(4):105–113
Titus VK, McDougall PP, Slevin DP (2010). Utilizing strategy formation mode and strategic learning to traverse various landscapes. Paper presented at the Academy of Management Conference, Montreal
van Woerkom M, Croon M (2009) The relationships between team learning activities and team performance. Pers Rev 38(5):560–577
Walter J, Kellermanns FW, Lechner C (2012) Decision making within and between organizations: rationality, politics, and alliance performance. J Manag 38(5):1582–1610
Wheelwright SC, Clark KB (1992) Revolutionizing product development: Quantum leaps in speed, efficiency, and quality. Simon and Schuster, New York
Wolf C, Floyd SW (2017) Strategic planning research: toward a theory-driven agenda. J Manag 43(6):1754–1788
Wu W-Y, Amaya Rivas A, Chen Y-C (2019) The role of team reflexivity as a mediator between project management skills, task familiarity, procedural justice, and product performance. J Manag Organ 25(6):876–895
Yousfi H (2019) Reshaping state/local communities relations in Tunisia: the socio-cultural and institutional challenges of the decentralization project. Eur Manag J 37(5):625–636
Yu ATW, Shen Q, Kelly J, Hunter K (2007) An empirical study of the variables affecting construction project briefing/architectural programming. Int J Proj Manag 25(2):198–212

Said Elbanna is a Research Professor of Strategy & Director of Center for Entrepreneurship & Organizational Excellence at Qatar University. He boasts an expansive tenure as a strategic management consultant, advisor, and educator. His research focuses on the impact of scholarly management research on organizational and societal development. He has authored or co-authored over 90 journal articles and book chapters on these subjects. His work has been published in top-tier international journals such as Strategic Management Journal and Long-Range Planning. Professor Elbanna has received numerous regional and international awards for his contributions to research, teaching, and service.

Mustafa Çolak is an Associate Professor of Strategic Management and the Head of The Department of Business Administration at the Social Sciences University of Ankara. He holds a dual MBA in Organization and Human Resource Management and Financial Management from the Bennett S. Le Bow College of Business at Drexel University, USA, and a PhD in Strategic Management with a focus on International Business from the Fox School of Business at Temple University, USA. His work has been published in such scholarly journals as R&D Management, European Journal of Innovation Management, and Journal of General Management. His current research interests include strategic planning, strategy implementation, and innovation in SMEs.

Chapter 12
Hedging Ratio Between Bitcoin and Altcoins Calculation and Creating an Optimal Portfolio

Fettah Kaba

Abstract Multivariate GARCH models were employed to examine the yield and volatility spreads among Bitcoin, Ripple, and Litecoin. Among these models, the VAR (1) model was identified as the most suitable within the BEKK-GARCH framework. The analysis revealed a unidirectional yield spread from Ripple to both Bitcoin and Litecoin. Additionally, volatility shocks originating from Ripple and Litecoin were found to affect Bitcoin. Continuing the research, the hedging ratio and optimal portfolio weights for bilateral combinations were established for Bitcoin, Ripple, and Litecoin. To hedge against a $1 long position in Bitcoin, it is necessary to take a short position of $0.56 in Ripple or $0.55 in Litecoin. Conversely, to hedge the risk associated with a $1 long position in Ripple, a short position of $0.70 in Litecoin should be taken. For portfolio construction, a $1 portfolio comprising Bitcoin and Ripple should allocate $0.77 to Bitcoin and $0.23 to Ripple. If the portfolio includes Bitcoin and Litecoin, then $0.94 should be invested in Bitcoin and $0.06 in Litecoin.

Keywords Bitcoin · Optimal portfolio · Cryptocurrency · Investor · Vector Autoregressive Model

This book chapter is derived from the thesis "Portfolio modeling with cryptocurrency return", which the author prepared with advisor Prof. Dr. Engin Demirel at Trakya University, Institute of Social Sciences, Department of Business Administration, Accounting and Finance. Advisor Prof. Dr. Engin Demirel has renounced his authorship rights.

F. Kaba (✉)
Vocational School of Social Sciences, Kırklareli University, Kırklareli, Turkiye
e-mail: fettahkaba@klu.edu.tr

K. T. Çalıyurt (ed.), *New Approaches to CSR, Sustainability and Accountability, Volume VI*, Accounting, Finance, Sustainability, Governance & Fraud: Theory and Application, https://doi.org/10.1007/978-981-96-5937-1_12

12.1 Introduction

Cryptocurrencies, unlike most existing financial assets, are not tied to any central authority and do not have a representative. Consequently, cryptocurrencies operate as a peer-to-peer electronic system, facilitating the easy transfer of online payments from one party to another (Corbet et al., 2019). Following the introduction of Bitcoin, the first cryptocurrency, numerous other cryptocurrencies have emerged in the market. The high volatility in cryptocurrency prices has captured the attention of investors. After Bitcoin, many altcoins have been introduced. Notably, some altcoins that incorporated technological elements and concepts from Bitcoin demonstrated remarkable performance in 2017. During this period, while Bitcoin yielded a return of 1300%, altcoins such as Litecoin and Ripple achieved returns between 5000% and 36,000% (Ji et al., 2019).

Since its launch, Bitcoin has attracted the attention of not only investors but also the scientific community. Numerous studies on Bitcoin have appeared in the literature. A significant focus of these studies has been the investigation of cryptocurrency volatility. Katsiampa et al. (2019a, b) examined volatility spreads and conditional correlations among Bitcoin, Ethereum, and Litecoin. Their findings indicated a bidirectional shock transmission effect between Bitcoin-Ethereum and Bitcoin-Litecoin pairs, along with a unidirectional shock from Ethereum to Litecoin. Additionally, the volatility spread effect was found to be bidirectional across all cryptocurrency pairs, with the conditional correlation mostly positive but varying over time.

Zięba et al. (2019) explored the relationship between changes in Bitcoin's price and other cryptocurrencies using the Vector Autoregressive Model (VAR). Dyhrberg (2016) investigated the similarities between Bitcoin, gold, and the dollar using GARCH models. Koutmos (2018) studied the relationships among various cryptocurrencies, concluding that Bitcoin is the primary source of yield and volatility spreads affecting other high-volume cryptocurrencies. Kumar and Anandarao (2019) provided evidence of volatility spreads from Bitcoin to Ethereum (BTC to ETH) and from Bitcoin to Litecoin (BTC to LTC).

Tu and Xue (2019) investigated yield and volatility spreads between Bitcoin (BTC) and Litecoin (LTC) using Granger causality tests and the BEKK-GARCH model. Their study provided evidence that the yield and volatility spreads only moved in one direction before the fork, specifically from BTC to LTC. After the fork, the direction of shock transfer reversed.

Luu Duc Huynh (2019) analyzed data from 2015 to 2019 for five cryptocurrencies, including BTC, XRP, and LTC, focusing on the risks of spillover between them. The study employed VAR and SVAR Granger causality tests as well as Gauss and Student's-t copula models. The findings suggested that ETH is likely to have a more independent relationship compared to other coins, while BTC was influenced by other cryptocurrencies.

Canh et al. (2019) examined a range of cryptocurrencies, including Bitcoin, Litecoin, Ripple, Stellar, Monero, Dash, and Bytecoin. Their study provided evidence of structural breaks within the cryptocurrency market, noting that these breakouts typically spread from smaller cryptocurrencies to those with larger market capitalizations. Additionally, the study identified strong positive correlations in volatility spreads among cryptocurrencies. It was found that Litecoin has the highest correlation value with Bitcoin among the pairs examined.

The research focuses on Bitcoin (BTC), Ripple (XRP), and Litecoin (LTC) among various cryptocurrencies. BTC was selected due to its status as the first and highest-valued cryptocurrency, while XRP and LTC were chosen for being the oldest cryptocurrencies introduced after BTC. The first part of the research examines the yield and volatility spreads between BTC, XRP, and LTC using multivariate GARCH models. In this context, the following hypotheses will be tested:

H1: The price of Ripple can be predicted using Bitcoin returns.
H2: The price of Litecoin can be predicted using Bitcoin returns.
H3: Volatility can be predicted among the selected cryptocurrencies.

In the second part of the study, the conditional variances and covariances of the most appropriate GARCH model will be used to calculate the hedging ratios between cryptocurrencies, following the method of Kroner and Sultan (1993). Additionally, using the method from the study by Kroner and Ng (1998), the binary optimal portfolio weights between cryptocurrencies will be calculated. At this stage, the following hypotheses will be tested:

H4: The hedging ratio can be calculated between BTC, XRP, and LTC cryptocurrencies.
H5: Bitcoin, Ripple, and Litecoin can be used to create an optimal portfolio.

This study aims to provide insights for investors on hedging risk and earning returns, as well as to contribute to the literature on cryptocurrencies.

12.2 Literature Review

There are numerous studies in the literature on yield spreads, volatility spreads, hedging ratio calculations, and diversification involving cryptocurrencies. For instance, Yousaf and Ali (2020a, b) applied the VAR-AGARCH model to data from Bitcoin, Ethereum, and Litecoin to examine yield and volatility spreads between these cryptocurrencies before and during the COVID-19 period. They calculated the optimal weights of binary portfolios created with cryptocurrencies and estimated the hedging ratios of the variables.

Wang et al. (2020) explored the relationship between Bitcoin and other altcoins, demonstrating the impact of price shocks in BTC on the returns of other cryptocurrencies. Qiao et al. (2020) investigated the joint movement between Bitcoin and other cryptocurrencies, noting that Bitcoin's returns and volatility are significantly influenced by other cryptocurrencies. Baur and Dimpfl (2018) applied the asymmetric GARCH model to Bitcoin and altcoins, concluding that the volatility of cryptocurrencies increases following positive shocks. Koutmos (2018) investigated the return and volatility shocks of cryptocurrencies, providing evidence that Bitcoin predominantly contributes to the spread of yield and volatility across all cryptocurrencies. Kumar and Anandarao (2019) stated that there is a significant volatility spread among BTC, ETH, and LTC. Canh et al. (2019) examined Bitcoin, Litecoin, Ripple, Stellar, Monero, Dash, and Bytecoin, presenting information about structural breaks and volatility spreads in these cryptocurrencies. Katsiampa (2019) provided evidence of a bidirectional shock transfer between BTC, Ether, and LTC. Katsiampa et al. (2019a, b) further demonstrated that shocks in BTC are more persistent compared to other altcoins.

Ampountolas (2022) evaluated the volatility dynamics of Bitcoin, Ethereum, Litecoin, and Ripple, offering a comprehensive empirical review of comparable prediction model performance through a systematic analysis of cryptocurrency price changes.

Bouri et al. (2021) examined the volatility of leading cryptocurrencies using the DCC-GARCH model. Their findings provide evidence linking investor sentiment to cryptocurrency volatility, arguing that market volatility increases when investors are extremely unhappy. Hossain and Ismail (2021) predicted the relationship between Bitcoin and other currencies using GARCH models, demonstrating that the GARCH (3,3) model offered the best predictions.

Zhang et al. (2021) utilized the VAR-DCC-GARCH model and compared it with other models such as VAR-BEKK-GARCH to evaluate its effectiveness in capturing cryptocurrency volatility, providing insights into the risk management and forecasting capabilities of different models. Chu et al. (2017) discussed data from Bitcoin, Dash, Litecoin, MaidSafeCoin, Monero, Dogecoin, and Ripple using GARCH models. Their findings indicate that the IGARCH and GJRGARCH models are the most effective, revealing extreme volatility in the daily prices of cryptocurrencies.

Aggarwal (2021) examined the volatility spread of Litecoin, Ripple, and Stellar. The study also calculated dual portfolio weights and hedging ratios, providing evidence of a volatility spread from Litecoin to other cryptocurrencies. Aggarwal suggested that investors hedge the risk of their long positions in Ripple by taking short positions in Stellar. W. Liu (2019) investigated the investment and diversification features of cryptocurrencies. The research findings indicate that adding cryptocurrencies to a portfolio significantly increases the Sharpe ratio. Dangi (2022) examined the hedging effectiveness of Bitcoin, Ethereum, Bitcoin Cash, and Ripple. The study found that the persistence of volatility spread is high for all cryptocurrencies except Ripple. Dangi recommended that investors with long positions in the spot market for Bitcoin, Ethereum, and Bitcoin Cash should hedge by taking short positions in the futures contracts of these respective cryptocurrencies. Qureshi et al. (2020) studied the interdependencies between Bitcoin, Ripple, Litecoin, Ethereum, and Bitcoin Cash using Wavelet Transform models. The analysis of the cross-wavelet transform showed that the contagion effect spread from Ripple and Ethereum to other cryptocurrencies.

12.3 Methodology and Data Set

12.3.1 Methodology

It is common to use GARCH models to predict volatility in financial markets. In this study, volatility modeling will be conducted using BEKK, CCC, and DCC models from the family of multivariate GARCH models.

12.3.1.1 BEKK-GARCH Model

Considering the multivariate conditional variance formulations developed by Engle and Kroner (1995) using GARCH (1,1), the conditional variance of the multivariate GARCH (1,1) model can be written as follows:

$$H_t = C_0' C_0 + \sum_{k=1}^{k} \sum_{j=1}^{q} A_{kj}' \varepsilon_{t-j} \varepsilon_{t-j}' A_{kj} + \sum_{k=1}^{k} \sum_{i=1}^{p} B_{ki}' H_{t-i} B_{ki}$$

In the equation, H_t represents the conditional variance of the multivariate BEKK-GARCH model. The term C in the model is an N × N dimensional matrix and is structured as an upper triangular matrix. In addition to the *CC* matrix, the diagonal coefficient matrices *AA* and B*B* are also N × N in size. When expanding the matrix of error terms, the BEKK-GARCH model can be expressed as follows:

$$H_t = C_0' C_0 + A_1' \varepsilon_{t-1} \varepsilon_{t-1}' A_1 + B_1' H_{t-1} B_1$$

12.3.1.2 DCC-GARCH Model

To investigate whether there are time-varying volatilities and correlations between the yield series of annuity-yielding cryptocurrencies, the DCC-GARCH model proposed by Engle (2002) was used. The DCC model consists of two stages. In the first stage, the GARCH parameters are estimated, and in the second stage, the correlations are estimated. The DCC conditional model of variance is expressed as follows:

$$H = D_t R_t D_t$$

In the equation, *Dt* represents the diagonal matrix of conditional variances obtained from the univariate GARCH(1,1) process, while R_t is expressed as the conditional correlation matrix.

$$D_t = diag\left(h_{11t}^{1/2}, \ldots h_{33t}^{1/2}\right) R_t = diag\left(q_{11t}^{-1/2}, \ldots q_{33t}^{-1/2}\right) Q_t diag\left(q_{11t}^{-1/2}, \ldots q_{33t}^{-1/2}\right)$$

The symmetric in the formula Q_t is a positive matrix. Refers $Q-$ to the 3 × 3 unconditional correlation matrix of standardized ξ_{it} residuals.

$$Q_t = (1-\theta_1-\theta_2)Q - +\theta_1\xi_{t-1}\xi'_{I-1} + \theta_2 Q_{t-1}$$

θ_1 and is a positive number whose sum is less than one. Finally, the model's correlation estimator is: θ_2

$$\rho_{i,j,t} = \frac{q_{i,j,t}}{\sqrt{q_{i,i,t}q_{j,j,t}}}$$

Similar to the research conducted by Sadorsky (2012), the multivariate GARCH model in this study was estimated using the BFGS algorithm and the quasi-maximum likelihood estimation.

12.3.2 *Hedge Ratios and Optimal Portfolio Weights*

In the second part of the study, the selected multivariate GARCH model will be further analyzed. Portfolio weights will be determined using the conditional variances of the chosen GARCH model. The method proposed by Kroner and Ng (1998) will be employed to determine the optimal portfolio weights of two assets.

$$\omega_{ij,t} = \frac{h_{jj,t} - h_{ij,t}}{h_{ii,t} - 2h_{ij,t} + h_{jj,t}}$$

Where ω_{ij} represents the optimal weight of asset i in a portfolio of assets i and j_j. H_{ii} and H_{jj} denote the time-varying conditional variances of the return series of assets i and j, respectively. H_{ij} denotes the time-varying conditional covariance between the return series of assets i and j.

In this context, the optimal weight of the entity i ($\omega_{ij,t}$) is determined by the following equation:

$$\omega_{ij,t} = \{0, if\ \omega_{ij,t} \langle 0\, \omega_{ij,t}, if\ 0 \le \omega_{ij,t} \le 1 1, if\ \omega_{ij,t} \rangle 1$$

The optimal weight of variable j ($\omega_{ij,\,t}$) is calculated by ($1-\ \omega_{ij,\,t}$). Thus, optimal weights that minimize portfolio risk are determined without causing a change in expected return rates.

After determining the optimal portfolio weights, the study examined the hedging ratios that could result from a long position in any of the relevant cryptocurrencies. The equation proposed by Kroner and Sultan (1993) was used to calculate the hedging ratios.

$$ohr_{t,i} = \text{cov}\left(\Delta x_{t,i}, \Delta y_{t,i}\right) / \sigma^2_{\Delta yt,i}$$

The $\Delta x_{t,i}$ logarithmic return series of the variable in the formula, $\Delta y_{t,i}$ shows the logarithmic return series of the variable to be used to hedge the risk; $\sigma^2_{\Delta yt,i}$ shows the conditional variance of the first variable that varies over time.

12.3.3 Data Set

The study used popular cryptocurrencies Bitcoin (BTC), Ripple (XRP), and Litecoin (LTC). To examine the relationship between BTC, XRP, and LTC, the daily closing prices of these cryptocurrencies from 10/04/2020 to 13/04/2024 were obtained from Yahoo Finance. Using this data, a series of daily logarithmic returns for the variables were created:

$$r = \ln\left(P_t / P_{t-1}\right) * 100$$

It shows the logarithmic return of the cryptocurrency at time *t*. P_t is the closing price of the cryptocurrency at time *t*, and P_{t-1} is the closing price of the cryptocurrency the previous day.

The return series of BTC, XRP, and LTC consists of 1464 observations. The movements of these return series over time are presented in Chart 12.1.

When examining the graphs of the returns for BTC, XRP, and LTC, volatility clusters in the series are notable. This characteristic indicates that the series exhibit properties typical of financial time series.

12.3.4 Descriptive Statistics

Descriptive statistics of BTC, XRP, and LTC returns are presented in Table 12.1.

The return series of BTC and LTC exhibit a negative and left-skewed structure, whereas the return series of XRP display a positive and right-skewed structure. Additionally, the BTC, XRP, and LTC series have high kurtosis values, indicating that the series have a leptokurtic, or peaked, structure. The high kurtosis values also imply that the return series do not follow a normal distribution.

The correlation results of the returns of BTC, XRP, and LTC are presented in Table 12.2.

In the yield series, the highest correlation is observed between BTC and LTC cryptocurrencies. The correlation coefficients between the returns of BTC, XRP, and LTC are high and positive.

Testing stationarity in time series is very important. In this study, the Augmented Dickey–Fuller (ADF) test (Dickey & Fuller, 1979) and the Phillips–Perron (PP) test (Perron, 1988) were used. The unit root tests and results for the BTC, XRP, and LTC returns are presented in Table 12.3.

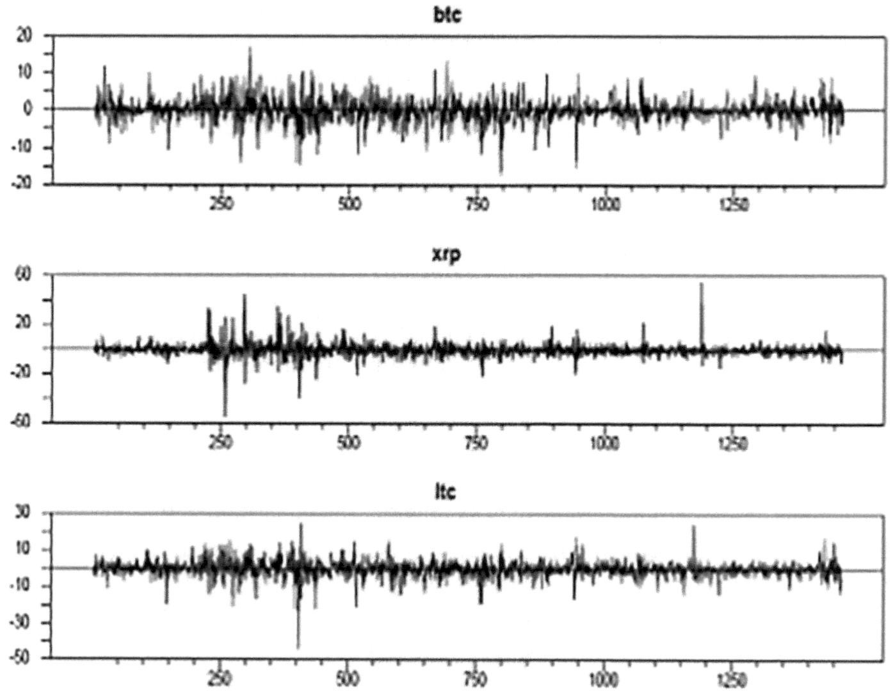

Chart 12.1 The movement of BTC, XRP, and LTC's returns over time

Table 12.1 Descriptive statistics on cryptocurrency returns

Indicator	BTC	XRP	LTC
Average	0.155912	0.072537	0.045107
Median	0.41276	−0.615387	−0.645858
Std. Deviation	3.279992	5.649467	4.754778
Skew	−0.180451	0.711025	−0.706978
Kurtosis	3.474470	21.290785	8.008531
Jarque-Bera	744.333***	27,774.505***	4034.286***

***, **, * 0.01, 0.05, and 0.10

The null hypotheses of the Augmented Dickey–Fuller (ADF) and Phillips–Perron (PP) unit root tests state that the series are not stationary. As shown in Table 12.3, both the ADF test null hypothesis and the PP test null hypothesis were rejected for the returns of BTC, XRP, and LTC. This indicates that there is no unit root in the return series of all three cryptocurrencies. The absence of a unit root in the return series implies that the series are stationary.

The ARCH test was conducted to determine if there is heteroscedasticity (varying variance) in the cryptocurrency returns used in the study. The results are presented in Table 12.4.

Table 12.2 Correlations of BTC, XRP, and LTC returns

	BTC	XRP	LTC
BTC	1	0.560	0.750
XRP	0.560	1	0.627
LTC	0.750	0.627	1

Table 12.3 Unit root test results of BTC, XRP, and LTC

	ADF		PP	
Variable	Constant	Fixed and trending	Constant	Fixed and trending
	t-Stat	*t*-Stat	*t*-Stat	*t*-Stat
BTC	−39.378***	−39.379***	−39.380***	−39.380***
XRP	−40.134***	−40.135***	−40.088***	−40.088***
LTC	−39.808***	−39.813***	−39.809***	−39.814***

***, **, * 0.01, 0.05, and 0.10

The null hypothesis of the ARCH test states that there is no ARCH effect in the series. Upon examining the ARCH test results for the return series of BTC, XRP, and LTC, the null hypothesis is rejected for all three-return series. This indicates that the ARCH effect is present in all yield series. Consequently, applying ARCH/GARCH models is more appropriate for analyzing the return series of these cryptocurrencies.

12.4 Empirical Findings

To predict the return and volatility of BTC, XRP, and LTC, an appropriate VAR (Vector Autoregressive) model was first established. The FPE and AIC criteria recommended a lag length of one (1) for the VAR model. In the estimation of the VAR model with a single lag, an autocorrelation test was performed on the residuals and no serial correlation was found in the model.

To make predictions with the VAR (Vector Autoregressive) model in financial time series, it is essential to examine the causality test results between the series. Therefore, the Granger Causality Test was conducted and the results are presented in Table 12.5.

According to the results of the Granger Causality Test in Table 12.5, there is a one-way Granger causality relationship between XRP and BTC, and between BTC and LTC return series, while there is a two-way Granger causality relationship between LTC and XRP. The Granger causality relationships in Table 12.5 align with the findings of Polat & Karakaya (2023). These results support the estimation of return spreads using the Vector Autoregressive (VAR) model with the return series of the existing cryptocurrencies.

In the financial time series analysis using the WinRATS 9 program, yield and volatility spread estimates were made using the VAR(1) BEKK-GARCH(1,1) and

Table 12.4 Varying variance test for the yield of BTC, XRP, and LTC

Variables	*F*-statistic
BTC	14.1323***
XRP	22.583***
LTC	6.928***

***, **, * 0.01, 0.05, and 0.10

Table 12.5 Granger causality test of BTC, XRP, and LTC yield

XRP ---> BTC	LTC ---> XRP	BTC ---> LTC	XRP ---> LTC
Coefficient	Coefficient	Coefficient	Coefficient
8.389***	5.993**	2.792*	5.223**

***, **, * 0.01, 0.05, and 0.10

VAR(1) DCC-GARCH(1,1) models with returns from Bitcoin, Ripple, and Litecoin. The student-t distribution was employed as the error distribution in the estimations. First, the estimation was performed using the VAR-BEKK-GARCH model, and the results are presented in Table 12.6.

Panel A provides the conditional mean estimates of the model. The impact of XRP's lagged return on BTC's current yield is negative and significant. Similarly, the impact of XRP's lagged return on XRP's current yield is also negative and significant. Additionally, the impact of LTC's current yield on itself and XRP's lagged yield is negative and significant. These results indicate that XRP is useful for predicting the returns of BTC, LTC, and itself.

In Panel B, the conditional variance-covariance estimates of the model are provided. The ARCH parameters $A(1{,}1)$, $A(2{,}2)$, and $A(3{,}3)$ represent the individual shocks of the variables. These three parameters are significant, indicating that the past shocks of the three cryptocurrencies have positively impacted the current conditional volatility. It is also observed that the individual ARCH parameters are smaller than the individual GARCH parameters $B(1{,}1)$, $B(2{,}2)$, and $B(3{,}3)$. This suggests that the long-term persistence of the variables is greater than their short-term persistence.

The parameter $A(1{,}2)$, which indicates the shock spread from XRP to BTC, is negative and significant. The shock spreads from LTC to BTC are also negative and significant. The $A(2{,}1)$ parameter, which indicates the shock spread from BTC to XRP, is insignificant, while the $A(3{,}1)$ parameter, which indicates the shock spread from BTC to LTC, is positive and statistically significant at the 10% level.

Examining the GARCH parameters $B(1{,}1)$, $B(2{,}2)$, and $B(3{,}3)$, which indicate the persistence of volatility, significant volatility spreads are observed for the three cryptocurrencies. BTC's GARCH parameter, $B(1{,}1)$, is the highest compared to the others, suggesting that BTC has the highest volatility persistence. Additionally, the one-way volatility spread from XRP to BTC is statistically significant at the 10% level. Similarly, there is a one-way volatility spread from LTC to BTC.

Robustness tests of the model are provided in Panel C. Upon examining the robustness tests, it was found that there is no autocorrelation issue in the residuals

Table 12.6 VAR-BEKK-GARCH model prediction results

	BTC	XRP	LTC
Panel A: Conditional average estimates			
Constant	0.135***	−0.133**	−0.040
BTC{1}	0.005	−0.001	0.012
XRP{1}	−0.051***	−0.103***	−0.047*
LTC{1}	−0.030	−0.029	−0.059*
Panel B: Conditional estimates of variance–covariance			
	Coefficient		Coefficient
C(1,1)	0.323**	A(3,1)	0.038*
C(2,1)	0.122	A(3,2)	0.017
C(2,2)	0.943***	A(3,3)	0.511***
C(3,1)	0.068	B(1,1)	0.974***
C(3,2)	0.624*	B(1,2)	0.063*
C(3,3)	1.230***	B(1,3)	0.147***
A(1,1)	0.232***	B(2,1)	0.006
A(1,2)	−0.237***	B(2,2)	0.893***
A(1,3)	−0.417***	B(2,3)	0.001
A(2,1)	−0.014	B(3,1)	−0.002
A(2,2)	0.461***	B(3,2)	0.012
A(2,3)	0.006	B(3,3)	0.849***
		Shape	2.647***
Panel C: Model robustness tests			
	Coefficient	*P*-value	
Log L	−10330.861		
AIC	14.173		
SIC	14.307		
$Q_1(20)$	25.991***	0.16611	
$Q_2(20)$	17.637***	0.61131	
$Q_3(20)$	17.750***	0.60384	
$Q_1^2(20)$	27.204***	0.12962	
$Q_2^2(20)$	0.998***	1.0000	
$Q_3^2(20)$	17.419***	0.62564	
Multivariate Q Statistic	49.46***	0.29971	
Multivariate ARCH Statistic	129.47***	0.99829	

***0.01, **0.05, *0.10 represent the test critical values
$Q_1(20)$, Q_2 (20), Q_3 (20) are autocorrelations of the standardized residuals of BTC, XRP, and LTC, respectively
$Q_1^2(20)$, Q_2^2 (20), Q_3^2 (20) are autocorrelations of the square of the standardized residuals of BTC, XRP, and LTC, respectively

of the model and variables. Additionally, it was observed that the ARCH effect did not persist in the residuals of the model. These results collectively support the accuracy of the VAR(1) BEKK-GARCH model established with BTC, XRP, and LTC yield series.

The investigation of the relationship between BTC, XRP, and LTC was looked at with the VAR-DCC-GARCH model. The VAR-DCC-GARCH estimation results are presented in Table 12.7.

In Panel A, the conditional mean estimates of the model are presented. Notably, the impact of XRP's lagged return on BTC's current yield is found to be negative and statistically significant. Similarly, the impact of XRP's lagged return on its own current yield is also negative and significant. Furthermore, the impact of LTC's

Table 12.7 VAR-DCC-GARCH model prediction results

	BTC	XRP	LTC
Panel A: Conditional average estimates			
Constant	0.091**	−0.185***	−0.104**
BTC{1}	0.006	−0.006	0.007
XRP{1}	−0.054***	−0.103***	−0.042**
LTC{1}	−0.031	−0.029	−0.067
Panel B: Conditional estimates of variance–covariance			
	Coefficient		Coefficient
C(1)	0.257*	A(3,1)	0.185***
C(2)	0.801***	A(3,2)	0.015**
C(3)	1.634***	A(3,3)	0.131***
A(1,1)	0.211***	B(1)	0.878***
A(1,2)	0.000	B(2)	0.783***
A(1,3)	0.003	B(3)	0.815***
A(2,1)	0.109	DCC(A)	0.062***
A(2,2)	0.221***	DCC(B)	0.928***
A(2,3)	0.043	Shape	2.610***
Panel C: Model robustness tests			
	Coefficient	*P*-value	
Log L	−10345.61		
AIC	14.184		
SIC	14.292		
Q1(20)	29.849**	0.07235	
Q2(20)	15.904***	0.72256	
Q3(20)	20.275***	0.44088	
$Q_1^2(20)$	1.483***	0.14267	
$Q_2^2(20)$	0.580***	0.58013	
$Q_3^2(20)$	13.980***	0.83153	
Multivariate Q Statistic	50.897***	0.25271	
Multivariate ARCH Statistic	109.12***	0.99999	

***0.01, **0.05, *0.10 represent the test critical values

$Q_1(20)$, Q_2 (20), Q_3 (20) are autocorrelations of the standardized residuals of BTC, XRP, and LTC, respectively

$Q_1^2(20)$, Q_2^2 (20), Q_3^2 (20) are autocorrelations of the square of the standardized residuals of BTC, XRP, and LTC, respectively

current yield and XRP's lagged yield is negative and significant. These results suggest that XRP is indeed useful for predicting the returns of BTC, LTC, and itself.

In Panel B, the conditional variance-covariance estimates of the model are provided. Notably, parameters $A(1,1)$ $A(2,2)$, and $A(3,3)$ representing the individual shocks of the variables are found to be significant. Additionally, short-term shock spreads from BTC to LTC and from XRP to LTC are significant. Furthermore, the GARCH parameters, indicating the persistence of shocks for BTC, XRP, and LTC, are also significant. It is noteworthy that BTC exhibits higher volatility shock persistence compared to XRP and LTC.

The sum of the coefficients of the DCC(A) and DCC(B) parameters is found to be positive and statistically significant at the 1% level. Notably, the sum of these coefficients is less than one, indicating a dynamic where conditional correlations revert to the mean (Sadorsky, 2012).

In Panel C, upon examination of the robustness tests of the model, it was found that there was no autocorrelation issue in the variables and residuals of the model. Furthermore, it was observed that the ARCH effect did not persist in the residuals of the model. These results provide evidence supporting the validity of the VAR(1) DCC-GARCH model established with BTC, XRP, and LTC yield series.

The yield and volatility spreads of the BTC, XRP, and LTC return series were estimated using multivariate autoregressive models VAR-BEKK-GARCH and VAR-DCC-GARCH. It was observed that there were no autocorrelation issues in the residuals of all three models, and the ARCH effect did not persist. Subsequently, in the continuation of the study, model outputs with low Akaike Information Criterion (AIC) scores will be further analyzed. For the calculation of portfolio hedging ratios and optimal portfolio weights of BTC, XRP, and LTC cryptocurrencies, the outputs of the VAR(1) BEKK-GARCH model will be utilized. Specifically, the hedging ratios will be calculated based on the methodology proposed by Kroner and Sultan (1993). Table 12.8. presents the hedging ratios of BTC, XRP, and LTC cryptocurrencies.

Analyzing the hedging ratios between BTC, XRP, and LTC, it is determined that to hedge the risk of a \$1 long position in BTC, a short position of 56 cents in XRP is required. Similarly to hedge the risk of a \$1 BTC long position, a short position of 55 cents in LTC is necessary. Furthermore, to hedge the risk of a \$1 long position in XRP, a short position of 70 cents in LTC is advisable.

Table 12.8 Hedging ratios between BTC, XRP, and LTC

	Mean	ST Dev	Min	Max
BTC/XRP	0.56	0.19	0.02	0.98
BTC/LTC	0.55	0.13	−0.05	0.81
XRP/BTC	0.89	0.28	0.1	2.31
XRP/LTC	0.7	0.21	0.04	2.22
LTC/BTC	1.03	0.2	−0.08	2.13
LTC/XRP	0.82	0.24	0.01	1.61

In the final part of the study, the hypothesis "H5: Bitcoin, Ripple and Litecoin cryptocurrencies can be used to create an optimal portfolio"was tested. Using the equation proposed by Kroner & Ng (1998), binary optimal portfolio weights consisting of BTC, XRP, and LTC were created. Table 12.9 presents the weights of the binary cryptocurrency pairs in the portfolio.

Research findings indicate the following optimal portfolio weights:

- In a $1 dual portfolio consisting of BTC and XRP, 77 cents should be invested in BTC and 23 cents should be invested in XRP.
- In a $1 dual portfolio consisting of BTC and LTC, 94 cents should be invested in BTC and 6 cents should be invested in LTC.
- In a $1 dual portfolio consisting of XRP and LTC, 67 cents should be allocated to XRP, while 33 cents should be allocated to LTC.

12.5 Conclusion

In the study, the yield and volatility spreads of popular cryptocurrencies Bitcoin, Ripple, and Litecoin (BTC, XRP, and LTC) were first examined using the return series from 11/04/2020 to 13/04/2024. Subsequently, the hedging ratios for BTC, XRP, and LTC and their optimal weights in binary portfolios were calculated. Initially, the BEKK-GARCH and DCC-GARCH models were used to calculate conditional correlations. The results of all three models passed the robustness tests. However, for this study, the results of the BEKK-GARCH model, which had the lowest Akaike Information Criterion (AIC) were considered.

The current returns of the three cryptocurrencies were negatively and significantly impacted by XRP's return from the previous day. These findings indicate that the return of XRP can be used to predict the returns of BTC and LTC. The short-term volatility shock spread from XRP to BTC is negative and significant, as are the short-term volatility shock spreads from LTC to BTC. This suggests that an increase in LTC's volatility leads to a decrease in BTC's volatility. Conversely, the short-term volatility shock spreads from BTC to LTC are positive and significant.

The GARCH parameters for BTC, XRP, and LTC were found to be significant and high, with BTC exhibiting the highest volatility persistence among the three cryptocurrencies. The study's findings indicate that BTC returns cannot be used to predict the prices of XRP and LTC. Therefore, the hypotheses "H1: Bitcoin returns can be used to predict the price of Ripple" and "H2: Litecoin price can be predicted using Bitcoin returns" are not accepted. However, there is evidence that the

Table 12.9 Optimal portfolio weights with BTC, XRP, and LTC cryptocurrencies

	Mean	St Dev	Min	Max
BTC/XRP	0.77	0.20	0.05	1.00
BTC/LTC	0.94	0.10	0.50	1.00
XRP/LTC	0.67	0.33	0.00	1.00

volatility shock spreads of XRP and LTC can be used to predict BTC's shock spread. Consequently, the hypothesis "H3: Volatility between selected cryptocurrencies can be predicted" is accepted.

Analyzing the hedging ratios between BTC, XRP, and LTC, it is found that a $1 long position in Bitcoin can be hedged by taking a short position of 56 cents in Ripple. Similarly, a $1 long position in Bitcoin can be hedged by taking a short position of 55 cents in Litecoin. The BTC/LTC pair has the lowest hedging ratio at 55 cents. These findings support the acceptance of the hypothesis "H4: The hedging ratio between BTC, XRP and LTC cryptocurrencies can be calculated." Finally, dual portfolio weights were calculated for Bitcoin, Ripple, and Litecoin. The results indicate that in a binary portfolio of $1 composed of Bitcoin and Ripple, 77 cents should be invested in Bitcoin and 23 cents should be invested in Ripple. In a dual portfolio of $1 consisting of Bitcoin and Litecoin, 94 cents should be invested in Bitcoin and 6 cents in Litecoin. For a $1 dual portfolio of Ripple and Litecoin, 67 cents should be allocated to Ripple, while 33 cents should be allocated to Litecoin. The study's findings are consistent with the research conducted by Brik et al. (2022) and Yousaf and Ali (2020a, b). These results support the acceptance of the hypothesis "H5: Bitcoin, Ripple and Litecoin cryptocurrencies can be used to create an optimal portfolio."

This study examined the relationship between Bitcoin and the altcoins Ripple and Litecoin. The research findings are expected to provide valuable insights for both scientists and investors. However, the study is limited by the inclusion of only Ripple and Litecoin in addition to Bitcoin. Future studies are recommended to incorporate a broader range of cryptocurrencies and to create portfolios consisting of more than just dual cryptocurrency pairs.

References

Aggarwal V (2021) Optimum investor portfolio allocation in new age digital assets. International Journal of Innovation Science 14(3/4):648–658. https://doi.org/10.1108/IJIS-10-2020-0237

Ampountolas A (2022) Cryptocurrencies intraday, high-frequency, volatility, spillover effects, using univariate and multivariate GARCH models. International Journal of Financial Studies *10*(3):Article 3. https://doi.org/10.3390/ijfs10030051

Baur DG, Dimpfl T (2018) Asymmetric volatility in cryptocurrencies. Econ Lett 173:148–151. https://doi.org/10.1016/j.econlet.2018.10.008

Bouri E, Gabauer D, Gupta R, Tiwari AK (2021) Volatility connectedness of major cryptocurrencies: the role of investor happiness. J Behav Exp Financ 30:100463. https://doi.org/10.1016/j.jbef.2021.100463

Brik H, El Ouakdi J, Ftiti Z (2022) Roles of stable versus nonstable cryptocurrencies in Bitcoin market dynamics. Res Int Bus Financ 62:101720. https://doi.org/10.1016/j.ribaf.2022.101720

Canh NP, Wongchoti U, Thanh SD, Thong NT (2019) Systematic risk in cryptocurrency market: evidence from DCC-MGARCH model. Financ Res Lett 29:90–100. https://doi.org/10.1016/j.frl.2019.03.011

Chu J, Chan S, Nadarajah S, Osterrieder J (2017) GARCH modelling of cryptocurrencies. Journal of Risk and Financial Management 10(4):Article 4. https://doi.org/10.3390/jrfm10040017

Corbet S, Lucey B, Urquhart A, Yarovaya L (2019) Cryptocurrencies as a financial asset: a systematic analysis. Int Rev Financ Anal 62:182–199. https://doi.org/10.1016/j.irfa.2018.09.003

Dangi V (2022) Estimation and effectiveness of optimal hedge ratios of cryptocurrencies based on static and dynamic methodologies. Vision, 09722629221135878. doi:https://doi.org/10.1177/09722629221135878

Dickey DA, Fuller WA (1979) Distribution of the estimators for autoregressive time series with a unit root. J Am Stat Assoc 74(366):427–431. https://doi.org/10.2307/2286348

Dyhrberg AH (2016) Bitcoin, gold and the dollar – a GARCH volatility analysis. Financ Res Lett 16:85–92. https://doi.org/10.1016/j.frl.2015.10.008

Engle R (2002) Dynamic conditional correlation. J Bus Econ Stat 20(3):339–350. https://doi.org/10.1198/073500102288618487

Engle RF, Kroner KF (1995) Multivariate simultaneous generalized arch. Economet Theor 11(1):122–150

Hossain MJ, Ismail MT (2021) Is there any influence of other cryptocurrencies on Bitcoin? Asian Academy Of Management Journal Of Accounting And Financ 17:125–152

Ji Q, Bouri E, Lau CKM, Roubaud D (2019) Dynamic connectedness and integration in cryptocurrency markets. Int Rev Financ Anal 63:257–272. https://doi.org/10.1016/j.irfa.2018.12.002

Katsiampa P (2019) Volatility co-movement between Bitcoin and Ether. Financ Res Lett 30:221–227. https://doi.org/10.1016/j.frl.2018.10.005

Katsiampa P, Corbet S, Lucey B (2019a) High frequency volatility co-movements in cryptocurrency markets. J Int Financ Mark Inst Money 62:35–52. https://doi.org/10.1016/j.intfin.2019.05.003

Katsiampa P, Corbet S, Lucey B (2019b) Volatility spillover effects in leading cryptocurrencies: a BEKK-MGARCH analysis. Financ Res Lett 29:68–74. https://doi.org/10.1016/j.frl.2019.03.009

Koutmos D (2018) Return and volatility spillovers among cryptocurrencies. Econ Lett 173:122–127. https://doi.org/10.1016/j.econlet.2018.10.004

Kroner KF, Ng VK (1998) Modeling asymmetric comovements of asset returns. Rev Financ Stud 11(4):817–844

Kroner KF, Sultan J (1993) Time-varying distributions and dynamic hedging with foreign currency futures. The Journal of Financial and Quantitative Analysis 28(4):535–551. https://doi.org/10.2307/2331164

Kumar AS, Anandarao S (2019) Volatility spillover in crypto-currency markets: some evidences from GARCH and wavelet analysis. Physica A: Statistical Mechanics and Its Applications 524:448–458. https://doi.org/10.1016/j.physa.2019.04.154

Liu W (2019) Portfolio diversification across cryptocurrencies. Financ Res Lett 29:200–205. https://doi.org/10.1016/j.frl.2018.07.010

Luu Duc Huynh, T. (2019). Spillover Risks on Cryptocurrency Markets: A Look from VAR-SVAR Granger Causality and Student's-t Copulas. Journal of Risk and Financial Management, 12(2), 52. https://doi.org/10.3390/jrfm12020052

Perron P (1988) Trends and random walks in macroeconomic time series: further evidence from a new approach. J Econ Dyn Control 12(2):297–332. https://doi.org/10.1016/0165-1889(88)90043-7

Polat M, Karakaya O (2023) Causality relationship between cryptocurrencies in terms of return and risk. Bartın University Journal of the Faculty of Economics and Administrative Sciences 14(27):1–17. https://doi.org/10.47129/bartiniibf.1169216

Qiao X, Zhu H, Hau L (2020) Time-frequency co-movement of cryptocurrency return and volatility: evidence from wavelet coherence analysis. Int Rev Financ Anal 71:101541. https://doi.org/10.1016/j.irfa.2020.101541

Qureshi S, Aftab M, Bouri E, Saeed T (2020) Dynamic interdependence of cryptocurrency markets: an analysis across time and frequency. Physica A: Statistical Mechanics and Its Applications 559:125077. https://doi.org/10.1016/j.physa.2020.125077

Sadorsky P (2012) Correlations and volatility spillovers between oil prices and the stock prices of clean energy and technology companies. Energy Econ 34(1):248–255. https://doi.org/10.1016/j.eneco.2011.03.006

Tu Z, Xue C (2019) Effect of bifurcation on the interaction between Bitcoin and Litecoin. Financ Res Lett 31st. https://doi.org/10.1016/j.frl.2018.12.010
Wang G-J, Ma X, Wu H (2020) Are stablecoins truly diversifiers, hedges, or safe havens against traditional cryptocurrencies as their name suggests? Res Int Bus Financ 54:101225. https://doi.org/10.1016/j.ribaf.2020.101225
Yousaf I, Ali S (2020a) Discovering interlinkages between major cryptocurrencies using high-frequency data: new evidence from COVID-19 pandemic. Financ Innov 6(1):45. https://doi.org/10.1186/s40854-020-00213-1
Yousaf I, Ali S (2020b) The COVID-19 outbreak and high frequency information transmission between major cryptocurrencies: evidence from the VAR-DCC-GARCH approach. Borsa Istanbul Rev 20:S1–S10. https://doi.org/10.1016/j.bir.2020.10.003
Zhang Y, Zhu P, Xu Y (2021). Has COVID-19 changed the hedge effectiveness of Bitcoin?. Front Public Health, 9th https://www.frontiersin.org/articles/10.3389/fpubh.2021.704900
Zięba D, Kokoszczyński R, Śledziewska K (2019) Shock transmission in the cryptocurrency market. Is Bitcoin the most influential? Int Rev Financ Anal 64: 102–125. doi:https://doi.org/10.1016/j.irfa.2019.04.009. https://finance.yahoo.com/quote/BTC-USD/history (Accessed on 14/04/2024) https://finance.yahoo.com/quote/XRP-USD/history (Accessed on 14/04/2024) https://finance.yahoo.com/quote/LTC-USD/history (Accessed on 14/04/2024)

Index

K. T. Çalıyurt (ed.), *New Approaches to CSR, Sustainability and Accountability, Volume VI*, Accounting, Finance, Sustainability, Governance & Fraud: Theory and Application, https://doi.org/10.1007/978-981-96-5937-1

MIX
Papier aus verantwortungsvollen Quellen
Paper from responsible sources
FSC® C105338

If you have any concerns about our products,
you can contact us on
ProductSafety@springernature.com

In case Publisher is established outside the EU,
the EU authorized representative is:
Springer Nature Customer Service Center GmbH
Europaplatz 3, 69115 Heidelberg, Germany

Printed by Libri Plureos GmbH
in Hamburg, Germany